92' ✓

Invitation to Mathematics

Invitation to Mathematics

Konrad Jacobs

Princeton University Press
Princeton, New Jersey

Library of Congress Cataloging-in-Publication Data
Jacobs, Konrad, 1928–
[Resultate. English]
Invitation to mathematics/by Konrad Jacobs.
    p.   cm.
Includes bibliographical references and index.
ISBN 0-691-08567-6 — ISBN 0-691-02528-2 (pbk.)
1. Mathematics. I. Title.
QA36.J3213 1992
510—dc20   91–19263

Originally published as *Resultate: Ideen und Entwicklungen in der Mathematik, Band 1, Proben mathematischen Denkens*, Vieweg Publishers, Braunschweig/Wiesbaden, 1987

This book has been composed in Linotron Times Roman

Princeton University Press books are printed on acid-free paper, and meet the guidelines for permanence and durability of the Committee on Production Guidelines for Book Longevity of the Council on Library Resources

Printed in the United States of America by Princeton University Press,
Princeton, New Jersey
        10 9 8 7 6 5 4 3 2 1
(Pbk.) 10 9 8 7 6 5 4 3 2 1

Contents

Foreword

THIS BOOK grew out of a course called ''Mathematics for Philosophers'' that I have taught several times at the Universität Erlangen-Nürnberg. In writing this book, I have also been able to profit from the experience I have gained as a teacher of teachers, an occasional public lecturer, and an author of survey papers for periodicals such as *Die Naturwissenschaften*. The work I have done so far has been published in German in two volumes that are titled *Resultate 1* and *Resultate 2*. The present book is an English version of *Resultate 1*. Apart from a few recent changes, the English version of the text that is printed here was produced by Jonathan Lewin of Kennesaw State College. I am much indebted to him. It was a great pleasure to discuss with him not only many details of the formulation of the topics being presented, but also some actual changes in the presentation of these topics. I was thus able to profit from his own experience as a teacher and as a textbook author. I feel, however, that I should take responsibility for any flaws that the reader may find in this text.

I this book, I address myself to readers with a knowledge of high school mathematics, and to undergraduate students at colleges and universities. I would be especially delighted if high school teachers would take this book as an instrument of general education that will enable them to widen and deepen their perspective of mathematics. In considering the role of this book, I was constantly confronted with the basic question,

What should the book include?   How should I approach each topic?

No doubt, these questions are the primary ones that must be faced by any author of a scientific work. In my particular case, the questions boil down to the following more specific questions:

What should an interested nonmathematician know about mathematics?   What are the proper ways of acquiring this knowledge?

On a very general level, I should like to formulate my answers as follows:

The layperson should become acquainted with those features of mathematics that relate to his or her own personal experience. In so doing, the layperson may confront mathematics in the following ways:

as a helpful tool in his or her own daily business.

as a bright phenomenon of cultural history

as a treasure trove of fundamental thoughts and ideas

Mathematics is not only a storehouse of knowable facts. Above all, it is a playground for well-trained activity. A book like this should therefore show a certain balance between its roles in reporting facts, proving results, and issuing invitations to the reader to become actively involved. Certainly, the layperson will not expect this book to transform him or her into a professional mathematician. Thus as the profession of people whose work touches our lives every day.

Mathematical facts should be reported in a visual and revealing fashion, free of clumsy terminology.

Proofs should, as a rule, be given in the form of plausible reasoning that displays their underlying arguments. However, the reader should also see a few fully fledged proofs as well.

Examples should, as a rule, be chosen in such a way that they also provide a nonmathematical view of the topic being discussed.

After some thought, I decided to attempt to fulfill these requirements in the following fashion:

As for the content of the book, I decided to make the following offerings:

specimens of mathematical reasoning
guided tours into various wings of the mathematical edifice, with emphasis on the structure, method, and the applicability of our science
tales from the history of mathematics
information about the present state of the art

As for the presentation, I decided on the following courses of action:

I would divide the book into two volumes, with emphasis on variety in the first volume, and a more systematic presentation in the second.

I would present Volume 1 without set-theoretical formalism, and with words and pictures rather than formulas, whenever possible.

I would integrate historical observations into the text and provide data on the lives of mathematicians whenever feasible.

I would cite papers and monographs with their year of publication so that the reader might trace the constant evolution of mathematics.

I would discuss a variety of obvious questions, provide interdisciplinary hints, and place emphasis on general viewpoints.

As for the bibliography, I would provide emphasis on texts that suit the nonmathematician. However, I would also provide access to some mathematical textbooks and even to some original papers.

Within the framework that I have just outlined, it is clear that I cannot offer an exhaustive choice of topics in this work. There are vast areas of mathematics that I am not competent to present, and there are many others that simply cannot be reasonably presented to the layperson at all. I therefore confined my choice to subjects about which I am able—being very generous to myself—to describe in a more or less responsible manner. For the present volume, I have tried to keep pure and applied mathematics in balance, and to give some hints about algorithms (even though computer science is not a topic in this book). I have also tried to focus a little on foundations; more of that in Volume 2. I have also attempted to reveal a few "prestigious objects" or "jewels" of mathematics—objects that could be on display in a "Mathematical Smithsonian" or even on a postage stamp. Among these, I include

> the five Platonic solids
> the Möbius strip
> the Cantor discontinuum
> the Peano curve
> the Alexander-Briggs-Reidemeister table of knots
> the plane ornaments
> Alexander's horned sphere
> Antoine's necklace.

This volume contains six chapters (I–VI), and Volume 2 (presently available only in the 1990 German edition) contains four more:

> VII. Logic
> VIII. Sets, functions, relations
> IX. Numbers and algebraic structures
> X. General views of mathematics (including a thorough sketch of its historical development)

Not every theorem, definition, and so on that appears in this book is numbered. Those that do are numbered in the traditional three figures fashion. For example, Theorem II.5.3 is in Chapter II §5. The index is generous. I wanted to make the finding of topics easier, and also to give some impression of the daily language of mathematicians.

I am deeply grateful to a number of my colleagues and friends who gave me valuable advice. Above all, I am grateful to Christian Thiel and Rudolf Kötter (Philosophical Seminar, Erlangen) and to Jonathan Lewin (Kennesaw State College). Walter Felscher (Tübingen) drew my attention to the fact that Du Bois-Reymond has some precedence over Cantor as to the discovery of the diagonal argument and discontinuum (see Felscher [1978/79]). My warmest thanks go to my secretary, Helga Zech, who typed the manuscript in both the German and the English versions. Finally, I wish to express my heartfelt thanks to the editors, Vieweg (Wiesbaden, W. Germany), and Princeton University Press for their fine efforts in all stages of the publication of this book.

Erlangen, October 1991

Invitation to Mathematics

Chapter I · Geometry

MATHEMATICS can be divided roughly into three main subdisciplines:

algebra   analysis   geometry.

Among these, geometry has been the principal representative of mathematics for the past two thousand years. Even today, geometry is particularly loved by mathematicians and nonmathematicians alike, perhaps because it is so visual, so universal, and so rigorous. Accordingly, we shall use geometry as our entranceway to mathematics in this book.

In the broadest sense, geometry is the investigation of spatial phenomena. Accordingly, it is most naturally arranged according to dimension:

Dimension 1:   lines and curves
Dimension 2:   surfaces
Dimension 3:   solids.

The problems of geometry may be approached at several distinct levels. For example, the sophisticated theory of cartography is a far cry from the simple geometry that would be used to describe the distance one might walk in the park. Qualitative statements like

Whether we leave Britain by land or sea we must cross the border somewhere

are matched by quantitative statements like

We fly over the Atlantic on a great circle route in order to save fuel.

Down-to-earth geometric constructions like the design of a sundial or a painter's perspective are in sharp contrast to statements of principle like

There are only five regular polyhedra: the tetrahedron, the cube, the octahedron, the dodecahedron, and the icosahedron (the Platonic solids).
There is no vertex of order five in a regular planar ornament.
There are precisely 230 crystal classes.
Squaring of a circle, trisection of an angle, and doubling of a cube cannot be performed exactly with ruler and compasses alone.
The parallel postulate cannot be deduced logically from the other axioms of Euclidean geometry.

In §1 we remind the reader of several theorems of Euclidean geometry. In §2 we study the possibility or impossibility of a variety of geometric constructions. In §3 we establish the correspondence between symmetry and group theory. In §4 we review some important ideas and results that have been obtained in connection with attempts to systematize, and make rigorous, geometry, with non-Euclidean geometry, projective geometry, analytic geometry, and Felix Klein's Erlanger Programm. We close this chapter with a short section, §5, that reports some developments in geometry. In Chapter V we shall treat one more geometric topic: topology. For both mathematicians and nonmathematicians the following books are especially recommended: Weyl [1952], Coxeter [1969], and Fejes Tóth [1965].

§1   Some Classical Theorems of Euclidean Geometry

Euclidean geometry is named after Eukleides of Alexandria, who lived around the year 300 B.C., and who codified in his *Elements* (Gr. *ta stoicheia*, 13 books) the mathematical knowledge—largely geometrical—that had been accumulated in the Mediterranean world until then, with the Pythagoreans as the most important contributors between 600 and 400 B.C.

The basic geometrical figures that occur in Euclidean geometry are

points, lines, planes, line segments, angles, triangles, circles.

Euclidean geometry is concerned with the measurement of lengths and angles, and focuses on constructions with rulers and compasses.

In this section we remind the reader of some fundamental theorems of Euclidean geometry, and their proofs.

1.1   The Circumference Angle and the Thales Circle Theorems

Circumference Angle Theorem. Every point C on the bold segment of the circle ''sees'' the points A and B under the same angle.

Sketch of Proof. We may write this angle as $\alpha + \beta$, where α, β, 2α, and 2β are as shown in Figure I.1.1. Note that we have used some well-known theorems in order to justify these labelings. The reader is invited to recall those theorems for himself. Since $2\alpha + 2\beta$ remains constant as C varies, so does $\alpha + \beta$. As an exercise one may also prove the theorem of the case in which C lies on the medium-bold part of the circle.

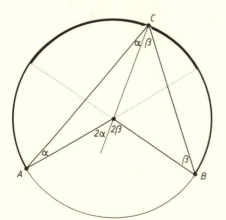

Figure I.1.1

Thales's Theorem. Every point C on a semicircle "sees" the diameter under a right angle. (A semicircle is also known as the Thales circle "over its diameter," after Thales of Milet who lived around 626–550 B.C.)

Proof. This result is the special case of the preceding theorem in which $2\alpha + 2\beta = 180°$.

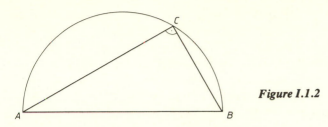

Figure I.1.2

1.2   *The Pythagorean Theorem*

This famous theorem is often considered the breakthrough discovery of the Greeks of around 550 B.C., and its proof is often considered to be the beginning of geometry as a rigorous science. However, as shown by Seidenberg [1962, 1978], the theorem and its proof had been common scientific knowledge as early as about 1000 B.C. in the cultural belt stretching from Europe to China. See also Waerden [1983].

Pythagorean Theorem. In a right triangle, the square of the hypotenuse equals the sum of the squares of the other two sides. We call the other two sides cathetes (see Fig. I.1.3).

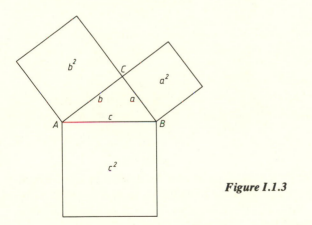

Figure I.1.3

Proof I. From Figure I.1.4. we see that

$$c^2 + 2ab = (a + b)^2.$$

Subtracting $2ab$ from both sides, we obtain $c^2 = a^2 + b^2$.

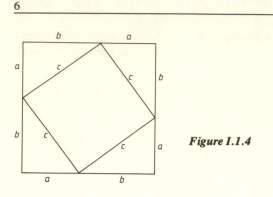

Figure I.1.4

Proof II. We may decompose the two squares on the legs into a total of 7 parts which fit exactly into the hypotenuse square (see Fig. I.1.5). The details are left to the reader.

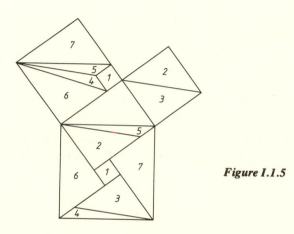

Figure I.1.5

There are many more proofs (see Loomis [1940], Lietzmann [1953]).

It has been conjectured that Egyptian surveyors made right angles by forming a 3:4:5 triangle from a rope with 13 knots. (Those surveyors were called *harpedonaptes* = rope spanners.) Since

$$5^2 = 3^2 + 4^2,$$

it follows from a converse of the Pythagorean theorem that the 3:4:5 triangle has a right triangle. Triples a, b, c of natural numbers that satisfy the equation $a^2 + b^2 = c^2$ are called *Pythagorean* triples. In addition to the Pythagorean triple 3, 4, 5, there are infinitely many others. What are they? Let us consider two odd numbers p and q with $p >$ q, and define $a = pq$, $b = \frac{1}{2}(p^2 - q^2)$, and $c = \frac{1}{2}(p^2 + q^2)$. We deduce easily that $c + b = p^2$ and $c - b = q^2$; and thus

$$c^2 - b^2 = (c + b)(c - b) = p^2q^2 = a^2,$$

i.e., $a^2 + b^2 = c^2$. This result tells us how to obtain infinitely many Pythagorean triples by taking infinitely many pairs p, q of odd numbers. It is easily shown—and the Pythagoreans already knew this—that all Pythagorean triples are obtained if a similar device with even numbers is added to the above method.

If we pose the same question with higher powers,

$$a^3 + b^3 = c^3, \qquad a^4 + b^4 = c^4, \qquad a^5 + b^5 = c^5, \ldots$$

we arrive at the famous *Fermat-Problem*. Pierre de Fermat (1601–1665) conjectured that the equation

(1) $a^p + b^p = c^p$

has no nontrivial solution in natural numbers a, b, c, and p if $p \geq 3$; in other words, he conjectured that this equation has no Diophantine solution.[1] His conjecture is known as the *Grand Fermat Conjecture* or *Fermat's Last Theorem (FLT)*, and he wrote it in the margin of a page of his personal copy of Diophantos. The original has been lost, but the book was reprinted together with Fermat's marginal remark in 1670. Fermat's remark reads as follows:

Cubum autem in duos cubos, aut quadrato-quadratum in duos quadrato-quadratos, et generaliter nullam in infinitum ultra quadratum potestatem in duas ejusdem nominis fas est dividere; cujus rei demonstrationem mirabilem sane detexi. Hanc marginis exiguitas non caperet.	It is impossible to decompose a third power into two third powers, a fourth power into two fourth powers, or, more generally, any power > 2 into two powers of the same order; for this I have discovered a truly remarkable proof, but this narrow page border cannot hold it.

One may show easily that Fermat's last theorem is equivalent to the assertion that equation (1) has no nontrivial solution in natural numbers a, b, and c where p is prime or twice a prime and $p \geq 3$. However, even today we still do not know whether there is an *infinity* of primes p for which equation (1) has no Diophantine solution. The state of the art as of 1976 is that equation (1) has no Diophantine solutions if p is any prime satisfying $3 \leq p \leq 125,000$.

A prize was offered for a proof of FLT by Paul Wolfskehl (1856–1906). This prize comes from stocks which are reportedly of some value even today. No one should dare to try to prove FLT without having made a thorough study of competent monographs on the subject, like Edwards [1977] or Ribenboim [1979]. At times "Fermatists" haunt mathematical institutes asking to have their "proofs" checked. Edmund Landau (1877–1938) is said to have responded to such requests with a form: ". . . has been checked by my assistant; the first mistake is on page . . ." Another mathematician simplified his task by sending one Fermatist's solution to another Fermatist. However, mathematicians should not underestimate the efforts of nonmathematicians. In principle, the layman has

---

[1] Diophantos of Alexandria lived around 250 A.D., and was especially interested in questions concerning the solvability of equations by integers and rationals. He left 13 books, *Arithmetika*, but of these, only 7 are still known to us today.

a chance of finding an original solution, precisely because he is not blinded by the trodden paths of old. But in the case of the FLT, the probability of stumbling upon an original idea is practically zero. See also Wagon [1986].

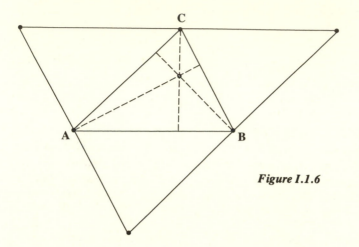

Figure I.1.6

1.3   The Altitudes Intersection Theorem

In an arbitrary triangle, the three altitudes intersect in one point. Quadruply the triangle *ABC* as shown. Everybody knows (you, too?) that the midpoint normals on the sides of a triangle meet in one point (namely, the center of its circumscribed circle). Apply this to the large triangle and realize that the heights of *ABC* coincide with the said normals. See Fig. I.1.6.

Note that an altitude of a triangle is a normal drawn from a vertex to the opposite side. The point of intersection of the altitudes is known as the orthocenter of the triangle.

1.4   Feuerbach's Nine-Point Circle

Karl Feuerbach (1800–1834) belonged to the famous Bavarian-Franconian genius family, which included criminologist Anselm (1775–1833), philosopher Ludwig (1804–1872, "God is a creature of man"), painter Anselm (1829–1880), and others (see Spoerri [1952]). Euclid should have been able to prove Feuerbach's theorem, which was published in 1822 (Feuerbach [1822]). Karl Feuerbach died insane (see Guggenbuhl [1955], which also proffers evidence that the theorem had previously been known to Brianchon and Ponçelet). The statement of **Feuerbach's theorem** is as follows:

Let *ABC* be any triangle. Then the following nine points lie on the same circle (called Feuerbach's Nine-Point Circle):

    (a) the intersection points H_A, H_B, and H_C of the three altitudes with the sides of the triangle,

    (b) the midpoints M_A, M_B, and M_C of the three sides of the triangle,

(c) the midpoints D_A, D_B, and D_C of the line segments running from the three vertices to the orthocenter of the triangle.

Figure I.1.7 shows, among other things, the diameter of a Thales circle and the medians of the given triangle. The task of constructing a proof of Feuerbach's theorem out of these hints is left to the reader (see Coxeter [1969], p. 18f).

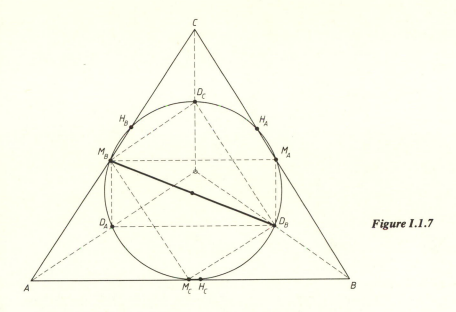

Figure I.1.7

1.5 The Regular Solids

In Euclid's time, the investigation of all lengths, angles, and so on that occur in the so-called Platonic solids was considered to be one of the foremost tasks of spatial geometry. In Figure I.1.8 we show five solids, along with their flat versions." Each of these five solids is regular in the sense that it has the same number of edges and faces at every vertex; all its edges have the same length; all its faces are regular n-gons, where n is 3, 4, or 5, and for each solid only one value of n occurs ($n = 3, 4, 5$); and, finally, the solid is convex. A primary question concerning regular solids is the following:

Are there regular solids other than the five shown in Figure I.1.8?

The answer is no, and the proof of this assertion was already known to Euclid. The proof is suggested by the easy fact that only 3-, 4-, and 5-gons can occur as faces of a regular solid; regular 6-gons can be fitted together only in a planar fashion.

If we weaken the condition of regularity, for example by omitting convexity, the number of possible solids increases sharply; see, for example, Coxeter [1969], Coxeter *et al.* [1982], and Fejes Tóth [1965]. Two famous not-quite-regular polyhedra are Kepler's "stella octangula" and the rhombic dodecahedron shown in Figure I.1.9.

As long as only five planets (aside from the Earth) were known (Mercury, Venus,

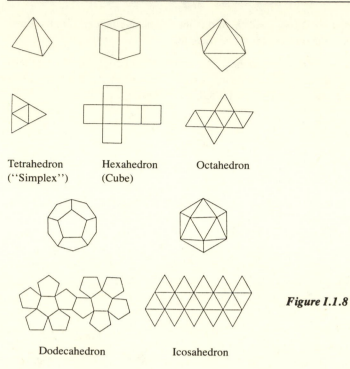

Tetrahedron   Hexahedron   Octahedron
("Simplex")   (Cube)

Dodecahedron   Icosahedron

Figure I.1.8

Mars, Jupiter, and Saturn), it was considered as plausible that this "five" should have something to do with the fact that there are only five regular solids. Johannes Kepler (1571–1630) made such speculations in Kepler [1619]. But this notion ended with the discovery of Uranus by Herschel in 1781, Neptune by Leverrier-Galle in 1846, and Pluto by Tombough in 1930. The philosopher Georg Wilhelm Friedrich Hegel (1770–1831) is said to have commented: "the worse for the planets." There are (so-called) esoterics who still take the idea seriously today.

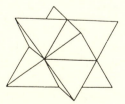

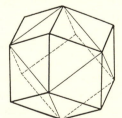

Figure I.1.9

Kepler's "stella octangula"   Rhombic dodecahedron

§2   *Possible and Impossible Constructions*

The geometric constructions that were prescribed by Euclid are those constructions that can be carried out using only ruler and compass. The important question that arises is what geometric figures can be constructed by these means. A systematic treatment of questions of this type may be found in Bieberbach [1952]. We report here on some of the classical results.

2.1   *Squaring the Circle*

A circle of radius 1 (a "unit circle") encloses a region with an area of π. This, in fact, is the *definition* of the number π. To square a given region means to construct a square whose area is the same as the area of the region. Figure I.2.1 allows us to visualize the problem of *squaring the circle*—in other words, the problem of using ruler and compasses to construct a square whose area is π.

In this form, the problem of squaring the circle withstood all the assaults that were made upon it over a period of more than two thousand years, but people still maintained hopes that it would be solved one day. Only in 1882 were these hopes buried once and for all by Ferdinand Lindemann (1852–1939), a student of Felix Klein's (1849–1925) and later himself the Doktorvater of David Hilbert (1862–1943). To understand the role of Lindemann's work in the problem of squaring the circle, we use the following simple fact: Rules-and-compasses constructions applied to a line segment of length 1 can produce only lengths and square areas that can be obtained from 1 by a finite number of applications of the four operations of arithmetic ($+$, $-$, $\cdot$, and $\div$) and the square root operation $\sqrt{}$. Numbers that can be obtained in this way are called *constructible* numbers, and it may be shown that all constructible numbers are *algebraic* numbers—in other words, all constructible numbers are roots of polynomials that have rational coefficients. The theorem that Lindemann proved in 1882 is that the number π is *transcendental* (Lindemann [1882]), in other words, that π is *not* algebraic, and Lindemann's theorem

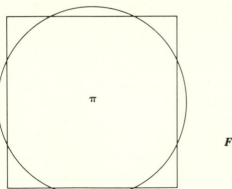

Figure I.2.1

therefore implies that it is impossible to construct a square whose area is π with ruler and compasses alone.

Squaring the circle became a synonym for "unsolvable problem" during the historical development of more than two thousand years. Dante (1265–1321) alluded to the problem in *Paradiso XXXIII*, 133–135, in connection with the mystery of trinity. The fact that $\pi = 3.1415926\ldots$ can be approximated by $22/7 = 3.1428571\ldots$ is said to have inspired the courtiers of Friedrich II von Hohenstaufen (1194–1250) to invent the poetical form of the sonnet 4:4:3:3, with 11 syllables in every line. There is also a rumor that this approximation induced Petrarca (1304–1374) to follow certain proportions in the composition of his book *Il Canzoniere* (Pötters [1983]).

Not all educated people who are familiar with the notion of geometric constructions seem to have heard of Lindemann's result. Thus, just as we have "Fermatists," we also have "circle squarers." To give them their due, these "circle squarers" have at least produced some clever approximate circle squarings. However, their approximations have little mathematical significance. A computer can be used to approximate π to as many decimal places as we like (subject only to the limitations of computer run-time and memory space), and using these approximations we can draw segments of rational length as close as we like to π. We remark in passing that although the decimal expansion of π can be calculated by a computer to a definite algorithm, it looks remarkably random, as may be guessed from

$$
\begin{array}{llllll}
\pi = 3.1415926535 & 8979323846 & 2643383279 & 5028841971 & 6939937510 \\
5820974944 & 5923078164 & 0628620899 & 8628034825 & 3421170679 \\
8214808651 & 3282306647 & 0938446095 & 5058223172 & 5359408128 \\
4811174502 & 8410270193 & 8521105559 & 6446229489 & 5493038196 \\
4428810975 & 6659334461 & 2847564823 & 3786783165 & 2712019091 \\
4564856692 & 3460348610 & 4543266482 & 1339360726 & 0249141273 \\
7245870066 & 0631558817 & 4881520920 & 9628292540 & 9171536436 \\
7892590360 & 0113305305 & 4882046652 & 1384146951 & 9415116094 \\
3305727036 & 5759591953 & 0921861173 & 8193261179 & 3105118548 \\
0744623799 & 6274956735 & 1885752724 & 8912279381 & 8301194912 \\
9833673362 & 4406566430 & 8602139494 & 6395224737 & 1907021798 \\
6094370277 & 0539217176 & 2931767523 & 8467481846 & 7669405132 \\
0005681271 & 4526356082 & 7785771342 & 7577896091 & 7363717872 \\
1468440901 & 2249534301 & 4654958537 & 1050792279 & 6892589235 \\
4201995611 & 2129021960 & 8640344181 & 5981362977 & 4771309960 \\
5187072113 & 4999999837 & 2978049951 & 0597317328 & 1609631859 \\
5024459455 & 3469083026 & 4252230825 & 3344685035 & 2619311881 \\
7101000313 & 7838752886 & 5875332083 & 8142061717 & 7669147303 \\
5982534904 & 2875546873 & 1159562863 & 8823537875 & 9375195778 \\
1857780532 & 1712268066 & 1300192787 & 6611195909 & 21644201989\ldots
\end{array}
$$

See Knuth [1968] and Wagon [1985]. It is therefore quite hard to memorize this decimal expansion. For a transcendental number whose decimal expansion is easy to memorize, we can look at 0.1 2 3 4 5 6 7 8 9 10 11 12 . . . (Mahler [1937]). More detailed infor-

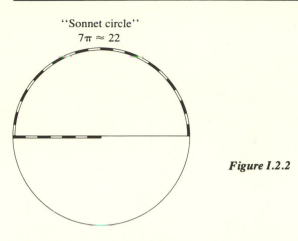

"Sonnet circle"
$7\pi \approx 22$

Figure I.2.2

mation about the number π may be found in Ebbinghaus *et al.* [1983] and Beckmann [1977].

2.2   *Construction of the Regular* n-*gon*

The precise statement of this problem is as follows: Using ruler and compasses only, subdivide a circle into n equal arcs. In other words, subdivide an angle of 360° into n equal angles. We have all learned to do this for a few values of n:

$n = 2$:   Halve the circle by drawing a diameter.

$n = 6$:   Carry the radius around the circle 6 times—it fits exactly.

$n = 3$:   Carry out the procedure for the case $n = 6$ and then omit every other point.

$n = 4$:   Draw two mutually perpendicular diameters.

Note that since it is easy to halve any given arc of circle, if we can subdivide a circle into n pieces, we can also subdivide it into $2n$ pieces. For example, we can handle the case $n = 12$.

The case $n = 5$ is more complicated, but it can be done. For thousands of years people wondered about the case $n = 7$, until young Carl Friedrich Gauss (1777–1855) proved on 30 March 1796 that it is impossible to perform a ruler-and-compasses construction of a regular heptagon (see Gauss [1796]). Gauss showed that the required lengths cannot be obtained from the number 1 because of their *algebraic* peculiarities. He was the first to provide an impossibility proof of this kind. The impossibility proof that Lindemann gave in 1882 for π was the last great achievement in this field. Gauss (see also Wantzel [1837]) proved the following statement:

If p is any prime number, then the regular p-gon can be constructed using ruler and compasses if and only if p can be represented in the form

$$p = 2^{2^m} + 1,$$

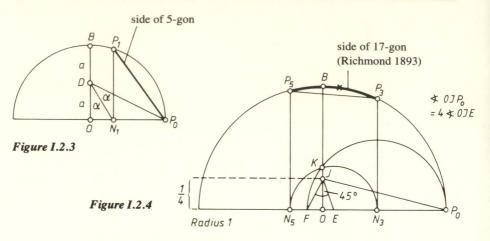

Figure I.2.3

Figure I.2.4

where m is a natural number.

The only known primes of this form are 3, 5, 17, 257, and 65537. In the course of time, all these constructions were actually carried out.

m	$p = 2^{2^m} + 1$	Construction by
0	3	folklore
1	5	antiquity
2	17	Gauss 1796
3	257	Richelot, Schwendenwein 1832
4	65537	Hermes 1879

Hermes's construction took about ten years. His work, which was more a calculation than a construction, was written out on large sheets of paper which are stored in a suitcase at the Mathematicians Institut Göttingen.

In a letter to Gerling dated 6 January 1819 (Werke X, 121–126), Gauss told the story of his discovery of the construction of the 17-gon:

> By strenuous pondering on the arithmetical connections among the roots I succeeded, during a vacation at Braunschweig, on the morning of the said day, while still in bed, to conceive these connections with utmost clarity, so that I could make the application to the 17-gon and the numerical confirmation right away.

He was proud of this discovery all his life. There is a legend that his tombstone shows a 17-gon, but this legend is untrue. However, the Gauss monument in Braunschweig displays a 17-point star.

Let us add one remark: Algebraically, our problem amounts to solving equations of the form $x^n = 1$; in other words, $x^n - 1 = 0$. Mathematicians therefore use the adjective "cyclotomic" when describing objects that relate to such equations.

2.3   *Trisection of Angles*

In 1837, Pierre Laurent Wantzel (1814–1849) proved that there is no ruler-and-compasses construction that can divide an arbitrary angle into three equal parts (Wantzel [1837]). The proof of this theorem is based upon the following two principles:

1. A given angle α can be constructed if and only if it is possible to construct a line segment of length $\cos \alpha$; and therefore from what we have said in §2.1 it follows that an angle α can be constructed if and only if the number $\cos \alpha$ can be constructed using the four operations of arithmetic, $+$, $-$, $\cdot$, and $\div$, and the square root operation $\sqrt{\ }$.

2. From the well-known trigonometric identity

   $$\cos 3\beta = 4 \cos^3\beta - 3\cos\beta$$

   it follows that for any angle α we have (putting $\alpha/3 = \beta$)

   $\cos \alpha = 4 \cos^3 (\alpha/3) - 3 \cos(\alpha/3)$. Now let α be any given angle and define $a = \cos\alpha$ and $x = \cos(\alpha/3)$. We can trisect the angle α if and only if the number $\cos(\alpha/3)$ is constructible, and therefore we can trisect the angle α if and only if the roots of the algebraic equation

(1) $4x^3 - 3x = a$

are constructible numbers.

It can be shown, however, that there is no general formula expressing the roots of (1) as constructible numbers, and therefore Wantzel's theorem follows. As stated above, Wantzel's theorem seems to exclude only a *single* method that will trisect an *arbitrary* angle. It leaves open the question of whether the set of angles can be partitioned into a number of separate classes such that for each class, a method of trisection exists that will work for every angle *in that class*. It can, however, be shown that there are simple special angles that cannot be trisected by ruler and compasses. The angle 60° is one of them; here we have $a = 1/2$ and in this case it can be shown that the roots of (1) are not constructible numbers. This observation also implies that the regular 9-gon is not constructible. See Bieberbach [1952] for details.

If we abandon the restriction of ruler and compasses, there do exist perfectly exact angle trisection methods. We depict one of these in Figure I.2.5. In this figure, the third

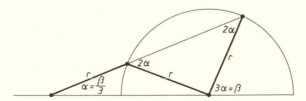

Figure I.2.5

appearance of the radius *r* on the left cannot be obtained by ruler and compasses; but it can be obtained quite well by sliding around a ruler that has markings on it.

If we are content with *approximate* trisections, then there are some clever ruler-and-compasses constructions. The celebrated mathematician Oskar Perron (1880–1975) didn't consider it below his dignity to publish a note "Eine neue Winkeldreiteilung des Schneidermeisters Kopf" (Perron [1933]). Needless to say, "angle trisectors" haunt the mathematical community like "Fermatists" and "circle squarers."

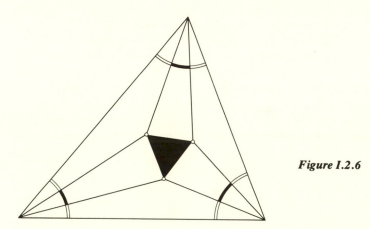

Figure I.2.6

The impossibility of angle trisection seems to have discouraged mathematicians from doing research on trisected angles. Thus the following theorem of Morley was discovered only around 1899:

Morley's Theorem. If we trisect the angles in an *arbitrary* triangle, then the respective dividing lines intersect in the vertices of an *equilateral* triangle.

See Fig. I.2.6 and Coxeter [1981], p. 24f.

2.4   Doubling the Cube

The problem of doubling the cube is said to have been posed by the priests of Apollo of Delos ("delic problem"), and is stated as follows:

> Given a cube with an edge of length 1 (a "unit cube"), construct the length of the edge of a cube that has twice the volume.

The algebraic counterpart to this problem is the problem of constructing the roots of the equation $x^3 - 2 = 0$ using $+$, $-$, $\cdot$, $\div$, and $\sqrt{}$. It can be shown that the number $\sqrt[3]{2}$ is not constructible, and thus the cube cannot be doubled using ruler and compasses (Wantzel [1837]).

2.5   The Question of Decomposition Equivalence

Two plane domains whose boundaries consist of straight-line segments (so-called polygonal domains) are said to be *decomposition equivalent* if the first domain can be decomposed into a finite number of triangles such that these same triangles, arranged differently, constitute a decomposition of the second domain. Decomposition-equivalent domains obviously have the same area. Note that the second proof that we gave of the Pythagorean theorem in §1.2 was based on decomposition equivalence. If we include the so-called Archimedean axiom among the axioms of plane Euclidean geometry, we can prove:

> Two plane polygonal domains have the same area if and only if they are decomposition equivalent.

Motivated by some remarks of Gauss, Hilbert [1900] posed as Problem Number 3 the question of whether a similar theorem holds true in three dimensions for polyhedra (using tetrahedra instead of triangles). For example, one might ask whether there exist two pyramids of the same base area and the same height (and therefore the same volume) that are not decomposition equivalent, or, alternatively, are not "complementation equivalent." Within a very short time, Hilbert's student Max Dehn (1878–1952) displayed two such pyramids (Dehn [1900, 1902]). Dehn used a well-established idea for the proof of nonequivalence: He defined an *invariant* with respect to equivalence, i.e., an assignment of numbers to pyramids such that the same value of the invariant is assigned to decomposition-equivalent pyramids. Then he constructed two pyramids of equal base area, and equal height, for which the value of the invariant was different. These two pyramids thus could not possibly be equivalent. Dehn's invariant was defined as a sum of edge angles in a decomposition of the given pyramid, neglecting integer multiples of 180°. If we decompose, then we obtain some inner edges with angle sum 360°—uninteresting; and we obtain some surface edges with angle sum 180°—uninteresting; but we also obtain some edges that lie on the edges of the original polyhedron, and here the interesting values arise which yield different values of the invariant for Dehn's two pyramids.

The question remained open whether equality of this and other similar invariants for two pyramids implies that they are decomposition (or "complementation") equivalent. Using a set of Dehn-like invariants, Sydler [1952/65] was able to prove such a theorem. His complicated proofs were greatly simplified by Jessen [1968].

§3   Groups of Rigid Motions

One of the most important developments in geometry was the interlacing of geometry with group theory, which came into being during the nineteenth century. In this bonding of two disciplines, each of them profited enormously. In this chapter, we shall focus on just two aspects of the symbiotic relationship between geometry and group theory. In the present section we are concerned with groups of rigid motions in the plane and in three-dimensional space, and with applications to the theory of ornaments and crystals. In §4 if this chapter, we shall show the role of group theory in attempts to systematize geometry (Felix Klein's Erlanger Programm of 1872). In addition to these two aspects of the role of group theory, there are many others. In fact, one may encounter groups nearly everywhere in mathematics. We shall see another example of the role of group theory when we discuss the so-called fundamental group of a topological space in Chapter V.

3.1   Congruence Motions of Geometric Figures

There are exactly six rigid motions in a given plane that will send a given equilateral triangle onto itself. We call these the *congruence motions* of the given triangle. See Figure I.3.1.

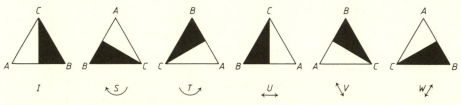

$$I \qquad \overset{\curvearrowleft}{S} \qquad \overset{\curvearrowright}{T} \qquad \underset{\longleftrightarrow}{U} \qquad \overset{\downarrow}{V} \qquad W$$

Figure I.3.1

Among these six motions we include the "motion" that doesn't actually move any points, but leaves them all where they were—in other words, the so-called identity mapping. It is a general methodological principle in mathematics not only to include such "trivial" cases, but even to pay special attention to them.

The first three motions shown in Figure I.3.1 are plane *rotations* around the center of the triangle, while each of the others consists of a *reflection* or "flipping." In the special case that we are considering, it is not important whether we see only the triangle itself as being moved in each motion, or whether our viewpoint is that the entire plane is moving and carrying the triangle along with it. It has, however, turned out that the latter viewpoint represents the more useful idea, especially when one extends the discussion to higher-dimensional spaces. We now list four simple observations about the six congruence motions of a given equilateral triangle:

(1) Two congruence motions D and E of the triangle can be combined into a congruence motion $E \circ D$ which operates on the triangle by applying the congruence motion D first, and then applying the motion E. Sometimes this method of combining motions is called concatenation.[2]

(2) The identity mapping I is a *neutral element* for the combination described in (1). In other words, if D is any congruence motion of the triangle, then

$$I \circ D = D = D \circ I.$$

(3) For every congruence motion D of the triangle, there is a unique congruence motion E such that

---

[2] Some authors combine D and E in the opposite order, but the order we have described is more common.

$$D \circ E = I = E \circ D.$$

This motion E is the motion that carries every point moved by D back to its original position. We call E the *inverse* of D and write it as D^{-1}:

$$D \circ D^{-1} = I = D^{-1} \circ D.$$

(4) The *associative law* holds. In other words, if D, E, and F are congruence motions of the triangle, then we have

$$D \circ (E \circ F) = (D \circ E) \circ F.$$

In this section we shall deal only with congruence motions of planar and three-dimensional figures. However, the four properties of congruence motions that we have described motivate the concept of an *abstract group*, and we take this opportunity to give the definition.

Definition of a Group: Let G be a set, and suppose that to every ordered pair (a,b) of elements of G there is associated an element $a \circ b$ of G. We call such an association $\circ$ a *composition* in G. Suppose that the following conditions hold:

Associative law: $a \circ (b \circ c) = (a \circ b) \circ c$ for all a, b, and c in G.

Neutral element: There is exactly one element e in G such that

$$a \circ e = a = e \circ a$$

for every element a of G.

Inverse: For every element a of G there is exactly one element a^{-1} of G such that

$$a^{-1} \circ a = a \circ a^{-1} = e.$$

The set G together with the composition $\circ$ is said to be a group. A group G, $\circ$ is said to be *commutative* if the following condition holds:

Commutative law: $a \circ b = b \circ a$ for all elements a and b of G.

Instead of saying "commutative" we sometimes say *Abelian* in honor of Niels Henrik Abel (1802–1829). Note that even when a given group is not commutative, we may have $a \circ b = b \circ a$ for some two given elements a and b of G. Two such elements are said to *commute*. The group itself is commutative when every two elements commute.

Finally, suppose that G, $\circ$ is a group, and that U is a subset of G. If the composition $a \circ b$ of any two elements a and b of U lies in U, and the inverse a^{-1} of any element a of U lies in U, then we call U (more precisely U, $\circ$) a *subgroup* of G (more precisely: of G, $\circ$).

We shall now return from our abstract excursion to the rigid motions of a given plane with the concatenation as their composition. We observe that

(1) The set of *all* rigid motions of the plane together with the composition $\circ$ is a group.

(2) The motions belonging to this group can be subdivided into two classes:
   (a) The class of all the motions that can be achieved just by combining translations and rotations of the plane.
   (b) The class of all the motions that cannot be achieved without a reflection or "flipping" about some line.

   The motions from class (a) are called *orientation-preserving* (or-pres), and those from class (b) are said to be *orientation-inverting* (or-inv).

(3) From the observations

   (or-pres) ∘ (or-pres) = (or-pres)
   $(\text{or-pres})^{-1}$ = (or-pres)
   the identity mapping I is or-pres

   we infer that the set of orientation-preserving motions of the plane form a subgroup of the group of all rigid motions of the plane.

(4) We observe that

   (or-pres) ∘ (or-inv) = (or-inv)
   (or-inv) ∘ (or-pres) = (or-inv)
   (or-inv) ∘ (or-inv) = (or-pres).

(5) Given any figure F in the plane, the set of all congruence motions of F is a subgroup G_F of the group of all rigid motions of the plane. We call this group G_F the *symmetry group* of F. The orientation-preserving motions in the group G_F form a subgroup of G_F.

(6) Suppose that F is a figure of the plane for which the group G_F is finite. Then the number of orientation-inverting motions in G_F is either 0 or the same as the number of orientation-preserving motions in G_F.

It is not hard to deduce (6) from (3), (4), and (5), but we shall not prove this result in its full generality. Instead, we shall look at the special case that we have already considered in which F is an equilateral triangle. In this case G_F consists of the six motions I, S, T, U, V, and W, which can be seen in Figure I.3.1. In this case the group G_F is usually denoted by D_3, and has the following *composition table*:

	I	S	T	U	V	W
I	I	S	T	U	V	W
S	S	T	I	V	W	U
T	T	I	S	W	U	V
U	U	W	V	I	T	S
V	V	U	W	S	I	T
W	W	V	U	T	S	I

From this table we may read, for example, that $I \circ I = I$, $V \circ T = W$, $S \circ V = W$, and so on. Within the table, in its upper left corner, we can see a smaller table that involves only the motions I, S, and T. From the symmetry of this smaller table about the diagonal that slopes down from left to right, we see that I, S, and T form an Abelian subgroup of D_3. I, S, and T are, by the way, precisely the orientation-preserving motions.

The fact that every row of the table contains every element of D_3 exactly once can be interpreted as saying that whenever a, b, and c belong to D_3 and $a \circ c = b \circ c$, then $a = b$. This property of D_3 is called the *right cancellation law*. By looking at columns instead of rows we can also see that D_3 has a left cancellation law.

We invite the reader to make similar investigations for the square, the regular pentagon, and other regular n-gons. You will find that the group of congruence motions of the regular n-gon is a group D_n that consists of $2n$ rigid motions of the plane, and that n of these are orientation-preserving and n are orientation-inverting. Note that the orientation-inverting ones are their own inverses because they are simply "flippings." We mention that from $n = 4$ onwards, the motions of the polygon do not produce all possible rearrangements of its vertices.

Much of what we have been saying in this section has to do with *symmetry*: the symmetry of figures, and also the symmetry that can be found in groups. In Figure I.3.2 we look at some other symmetry groups G_F for plane figures F, some of which even extend to infinity.

Generally, we think of figures F as being more symmetric the larger their groups G_F. In spite of the beautiful appearance of artistic ornaments, they reduce to a Spartan simplicity when we look only at their bare mathematical essentials.

We conclude this subsection by extending our investigation to three-dimensional figures. As for two-dimensional figures, the rigid motions of three-dimensional space form a group under the concatenation and fall into two classes:

> (a) Motions that can be achieved just by combining translations and rotations.
> (b) Motions that require some (mirror) reflection.

The motions of class (a) are called *orientation-preserving*, and form a subgroup of the group of all rigid motions of three-space. In every finite subgroup of this large group they either form the whole group or exactly half of it (a subgroup). The motions of class (b) are called *orientation-inverting*. When we looked at orientation-reversing motions in a plane, we would visualize them either as reflections or as flippings of the plane in a three-dimensional space containing it. In our three-dimensional case we can visualize such motions only as a reflection. On the other hand, if we work with coordinates, then we have no difficulty regarding these motions either as reflections or as flippings in a four-dimensional space (see subsection 4.3).

Just as we saw for motions in a plane, the concatenation of two orientation-preserving motions in three-space is orientation-preserving, the concatenation of an orientation-preserving motion with an orientation-inverting one is orientation-inverting, and so on. In order to determine whether or not a given motion in space is orientation-preserving, we let it act on "typically oriented figures" like hands, books, labeled tripods, or tetrahedra. See Figure I.3.3.

If we look, for example, at the symmetry group G_F of a regular tetrahedron F, then

F	G_F
Asymmetric frieze 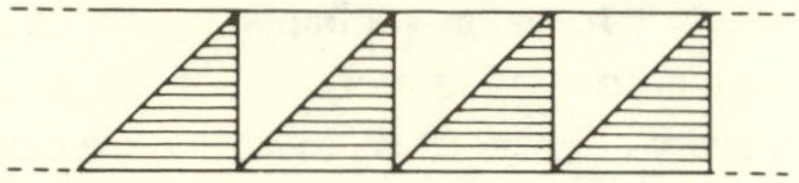	G_F consists of all n-fold repetitions of a translation by the period of the frieze (where n is any integer). G_F is composition-equivalent ("isomorphic") to the group Z of all integers with the usual addition $+$ as composition. In particular, G_F is Abelian.
Symmetric frieze 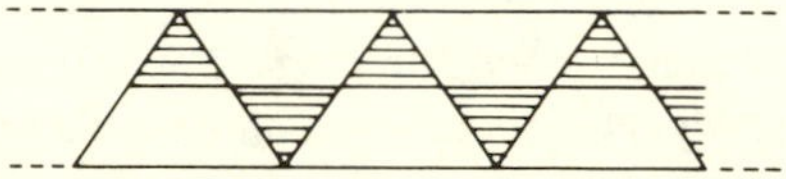	G_F contains translations, and also flippings about vertical lines, and combinations of flippings and translations that leave the midline of the frieze invariant (paddle motions). The group G_F is non-Abelian.
n-fold asymmetric rosette 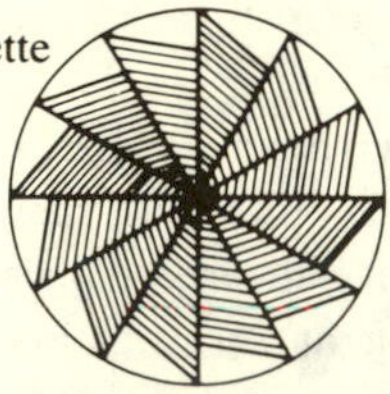	G_F contains only orientation-preserving rotations around the midpoint of the rosette, and is Abelian.
n-fold symmetric rosette 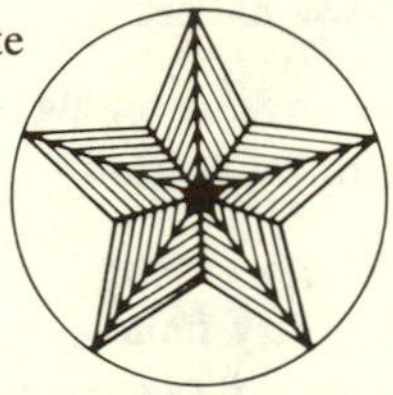	G_F contains both rotations and flippings and is non-Abelian.
Asymmetric tiling 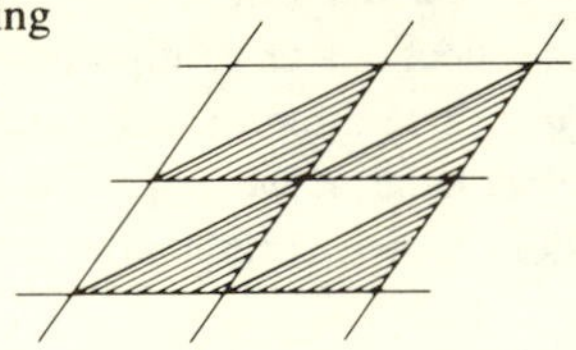	G_F consists of translations only and is Abelian.
Honeycomb tiling 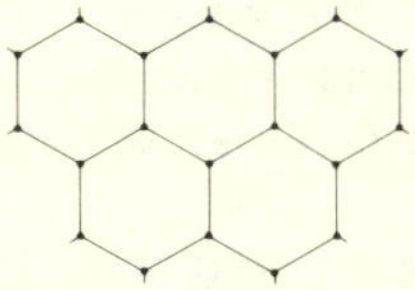	G_F contains translations and rotations and flippings, and is non-Abelian.

Figure I.3.2

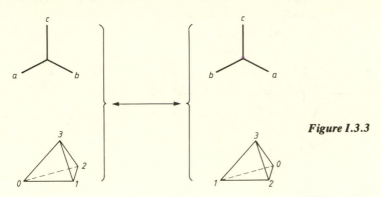

Figure I.3.3

we find the 24 rigid motions that are listed in Figure I.3.4. The left half of this figure shows the 12 orientation-preserving motions in this group, and the right half shows the orientation-inverting ones. The four lines in the figure correspond to the four possibilities of replacing the lower left vertex of the tetrahedron by itself, or by one of the three other vertices.

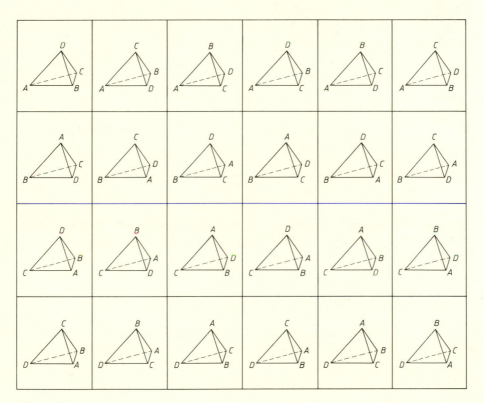

Figure I.3.4

Name	Figure	Number of congruent motions
Tetrahedron	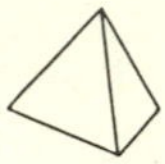	24
Cube	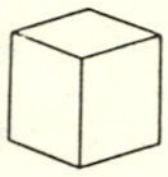	48
Octahedron	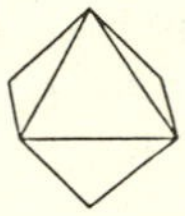	48
Dodecahedron	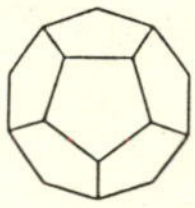	120
Icosahedron	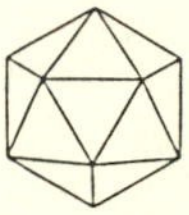	120

Figure I.3.5

There are analogous listings for the symmetry groups of the other Platonic solids, called the "Platonic groups." Figure I.3.5 gives their numbers of motions only. It's no wonder that we find the same numbers:

      48 for both the cube and the octahedron
      120 for both the dodecahedron and the icosahedron.

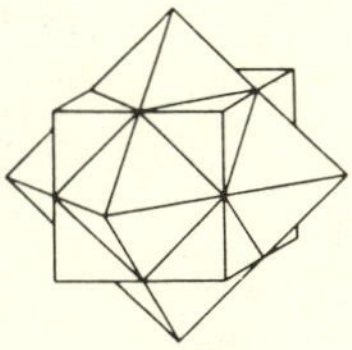

$$G_{cube} = G_{octahedron}$$

$$G_{dodecahedron} = G_{icosahedron}$$

Figure I.3.6

Actually the groups are identical, as can be seen by suitably fitting a cube into an octahedron, or by fitting a dodecahedron into an icosahedron. See Figure I.3.6.

In his famous book on the icosahedron (Klein [1884]), Felix Klein (1849–1925) put the icosahedron group into a huge network of algebraic and geometric correspondences. Some specialists consider this to be his most beautiful book. To Klein himself it meant resurrection after a nervous breakdown in which he believed he had lost his best creative powers.

3.2   The Classification of Symmetry Groups, Ornaments, and Crystals

We have defined the symmetry group G_F of a given figure F both in the plane and in three-space, and we have displayed some concrete examples of such groups. We have not, however, displayed them all. In addition to these examples there are many others, and the question arises whether it it possible to make a list of symmetry groups and say, "Here are all possible symmetry groups. No others will ever be found."

This type of question arises in mathematics over and over again. We form a mathematical concept by setting up its definition; then we look for examples of mathematical objects that fit this definition; and finally, we attempt to *classify* all the mathematical objects that fit the definition. We like to know when we have reached the point at which any mathematical object that fits the definition will not really be "new," but will be "isomorphic" to one of the objects we have already described. Roughly speaking, two objects are isomorphic when their structures are essentially the same, and they differ only in the notations that are used for them.

Formally, the question may be described as follows: We partition the set of all possible examples into classes in such a way that two examples are in the same class if and only if they are isomorphic (we may call these classes isomorphism classes). The question now reduces to the problem of drawing up a list that contains one and only one example from each class, or what mathematicians call a complete repetition-free list of representatives from the classes. This problem is called the *classification problem* for the given mathematical concept.

In our particular case, we are looking for a *classification of all planar (or spatial) symmetry groups*. We therefore need to eliminate all duplications from the list of all G_F's. For example, G_F is the same for a cube and an octahedron. We shall now simplify the problem a little by replacing the set of all G_F's by a somewhat smaller set.

Given a group G of rigid motions of (say) a plane, and given a point x in the plane, we define the G-orbit of x to be the set of all those points in the plane that can be obtained from x by applying the motions in the group G. We call such an orbit *discrete* if it has only a finite number of points in every bounded region of the plane. We call the group G *discrete* if the G-orbit of every point x is discrete. In the preceding examples, all the symmetry groups G_F were discrete. Note, however, that if F is a singleton x_0, then the G_F-orbits are circles with center x_0 and so G_F is not discrete. We shall exclude such cases and restrict our attention to those figures F for which the group G_F is discrete. In other words, we are concerned with the problem of *classifying the discrete groups of rigid motions of a plane or a three-dimensional space*.

In order to tackle this problem, we need an overview of the possible types of rigid motions. These are

In the plane	In three-space
identity	identity
rotations around a given point .	rotations around an axis
mirror reflections about straight lines	mirror reflections about planes
translations	translations
paddle motions	spiral motions

By paddle motion we mean a reflection about a line following a translation about that line. By spiral motion we mean a rotation around an axis following a translation parallel to that axis. The discrete groups may be classified according to the following scheme: For every discrete group G we obtain a fundamental domain by picking exactly one point from every G-orbit. This can be done in many ways. As a matter of fact, if we start with any one fundamental domain of a group G and act on it with a motion in G, then we obtain a second fundamental domain of G which is disjoint from the first one. The fundamental domains of G obtained this way cover the entire plane (or three-space) and we call them a *tiling* of the plane (or three-space). The desired classification is achieved by listing all possible tilings, and indicating how the rigid motions act on them. In the plane the result (Pólya [1924], here quoted from Fejes Tóth [1965]) looks as shown in Figure I.3.7.

The meaning of this tableau is the following:

With every figure we associate the tiling of the plane by fundamental domains, the "tiles"; each tile is given an "orientation" indicated by a curl. Now if we grasp any one of these tiles and move the plane in such a way that our specified tile comes to coincide with some other tile, including their orientations ("curl on curl"), then we have carried out one of the motions of the group G. All the motions of G can be obtained in this way. $\ominus$, $\oslash$, and $\oplus$ are code symbols for rotations by $\pm 180°$, $\pm 120°$, and $\pm 90°$, respectively.

This classification of all discrete planar groups is the branch of mathematics known as the *theory of ornaments*. This theory explains, among other things, why there cannot exist a planar ornament with perfect symmetry points of order five, thus solving a problem that had haunted the Islamic ornamentalists for centuries (see Critchlow [1976], El-Said-Parman [1976], and Rempel [1961]). In Figure I.3.8 we compare an Islamic ornament of "order five" with a modern mathematical one (Kline [1979]).

The theory of ornaments can be applied in much the same way to a three-dimensional space, where the classification problem is one of classifying "crystals" and is one of the great achievements of the nineteenth century. The three-dimensional theory leads to 230 different groups, the crystal groups. The original papers are Fedorov (1853–1919) [1891], Schoenflies (1853–1928) [1891], and Barlow (1845–1934) [1894]. A modern version of this work can be found in Fejes Tóth [1965]. For the algorithmic aspects of this theory one may consult Zassenhaus [1948]. If we confine ourselves to *finite* discrete groups, then the list contains only 32 groups, and these had already been listed by Hessel in 1830 (Hessel [1830, 1831]). Hessel's work is confined to those groups of motions that leave a particular point fixed, and these groups are known as the 32 *point groups*. These 32 point groups are not sufficient to describe the crystals that occur in nature. For this

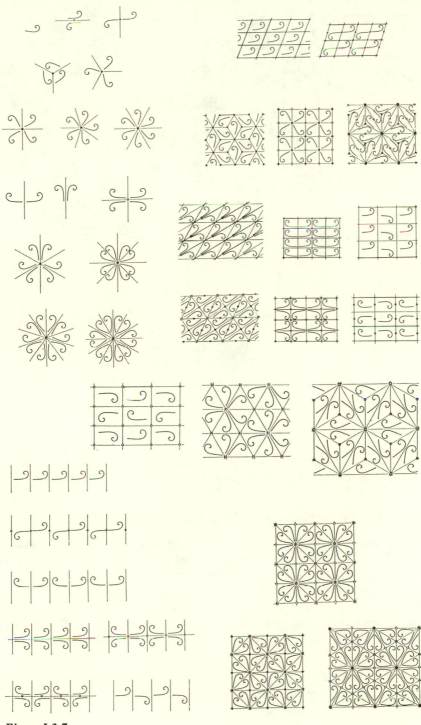

Figure I.3.7

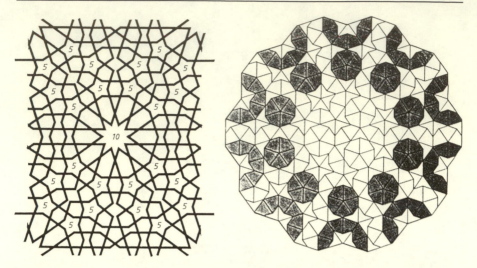

Figure I.3.8

purpose 47 groups are required, and these give us the list of 47 simple crystal shapes (see Shubnikov-Koptsik [1974]). Recently, the study of "four-dimensional crystals" has attracted some interest (Brown *et al.* [1978]). The idea of the lattice structure of crystals can be traced back to Christiaan Huygens (1629–1695) and Johannes Kepler (1571–1630).

§4   *Systematizing Geometry*

In this section we describe some mathematical enterprises whose aim is to bring geometry, or parts of it, into a rigorous system.

4.1   *The Axiomatic Edifice of Euclidean Geometry*

Already Euclid of Alexandria (around 300 B.C.) succeeded in representing geometry as a theory with certain systematic features in his thirteen-book *Elements* (Gr. *ta stoicheia*). He did not just set down construction after construction, theorem after theorem, but attempted to make the theory satisfy principles of "soundness" that had been proposed by such people as Aristotle. A sound theory should start with some "axiomata" (*koinai ennoiai*) and should proceed by definitions and proof. Thus Book I of the *Elements* begins with

> Definitions (Gr. *horoi*)
> α′. A point is something that has no parts.
> β′. A line is length without width.
>
> . . .

δ'. A straight line is a line that lies equally with respect to all its points.

ε'. A surface is something that has only length and width.

. . .

Today we consider these initial explanations to be unsatisfactory and imprecise. Euclid's development of the theory makes repeated use of intuitions that are allowed to sneak in because they are visually "obvious," but that have no clear logical status. The Aristotelian program was first carried out rigorously for Euclidean geometry by Moritz Pasch (1843–1930) [1882] and notably by David Hilbert (1862–1943). In his book *Grundlagen der Geometrie* (Hilbert [1899]) Hilbert started with the axioms displayed on page 30.

Hilbert's radical interpretation of Aristotle's demands is revealed in his introduction, which contains the following words:

> In formulating the axioms, it is of no importance with what imaginations we associate the names of concepts occurring in axioms. Only the logical relations established between these names in the axioms count. We may use visualizations and interpretations only in order to motivate the direction of our development, but these motivations should never enter the formal definitions, proofs, and theorems. The rigor of the development should rely on the laws of logic alone.

Allegedly, Hilbert once said that instead of "point," "straight line," and "plane," you could just as well write "chair," "table," and "beer mug." Hilbert's expulsion of physical content was soon criticized by the logician Friedrich Ludwig Gottlieb Frege (1848–1925). The objection to Hilbert's idea was that mathematics would not be a science any more since it would have no object of investigation. Only with the inclusion of a physical interpretation would this arbitrary formal puzzle become a science again. According to Frege, what Hilbert had written was only the bare form of a theory (Frege [1903], Nelson [1928]).

Practically all mathematicians nowadays share Hilbert's view, not only with regard to geometry, but with regard to all of mathematics. Thus Frege's objection applies to all of contemporary mathematics. In fact, I believe that mathematics, as we understand it nowadays, is not a science in the usual sense, but a meta-science. Although there are some mathematical disciplines that refer *a priori* to some particular interpretations that give them the status of a science in the usual sense, for most others this is not the case. For the latter, one may say: just add an interpretation and out comes a science. I, as a mathematician, do not find this seriously objectionable. It even distinguishes mathematics: this super-science represents the highest form of organization of knowledge, even more than the classical ideal of a science.

4.2   The Parallel Postulate and Non-Euclidean Geometry

Not until we have set up a theory that satisfies Hilbert's ideal can we investigate precisely the inner structure of the theory and give complete answers to questions like

What can we prove in this theory by what means, and what can we not prove?

Hilbert's Axioms of Euclidean Geometry (excerpts from the 1902 English translation of Hilbert [1899])

Group I. Axioms of Connection. The axioms of this group establish a connection between the concepts *point*, *straight line*, and *plane*. These axioms are as follows:

I.1. Two distinct points A and B always completely determine a straight line a. We write
$$AB = a \text{ or } BA = a.$$
Instead of "determine," we may also employ other forms of expression. For example, we may say A "lies upon" a; A "is a point of" a; a "goes through" A "and through" B; a "joins" A "and" or "with" B, etc. If A lies upon a and at the same time upon another straight line b, we also make use of the expression: "The straight lines" a "and" b "have the point A in common," etc.

I.2. Any two distinct points of a straight line completely determine that line; that is, if $AB = a$ and $AC = a$, where $B \neq C$, then also $BC = a$.

.   .   .   .   .

Group III. Axiom of Parallels (Euclid's Axiom). The introduction of this axiom greatly simplifies the fundamental principles of geometry and facilitates in no small degree its development. This axiom may be expressed as follows:

III.1. In a plane there can be drawn through any point A, lying outside of a straight line a, one and only one straight line that does not intersect the line a. This straight line is called the *parallel* to a through the given point A.

This statement of the axiom of parallels contains two assertions. The first of these is that in the plane a, there is always a straight line passing through A that does not intersect the given line a. The second states that *only one* such line is possible. The latter statement is the essential one.

Group II. Axioms of Order. The axioms of this group define the idea expressed by the word "between," and make possible, upon the basis of this idea, an order of sequence of the points upon a straight line, in a plane, and in space. The points of a straight line have a certain relation to one another which the word "between" serves to describe. The axioms of this group are as follows:

II.1. If A, B, C are points of a straight line and B lies between A and C, then B also lies between C and A.

II.2. If A and C are two points of a straight line, then there exists at least one point B lying between A and C and at least one point D so situated that C lies between A and D.

II.3. Of any three points situated on a straight line, there is always one and only one that lies between the other two.

.   .   .   .   .

Group IV. Axioms of Congruence. The axioms of this group define the idea of *congruence* or *displacement*.

Segments stand in a certain relation to one another, which is described by the word "congruent."

IV.1. If A, B are two points on a straight line a, and if A' is a point upon the same or another straight line a', then, upon a given side of A' of the straight line a', we can always find one and only one point B' so that the segment AB (or BA) is congruent to the segment $A'B'$. We indicate this relation by writing

$$AB \equiv A'B'.$$

Every segment is congruent to itself; that is, we always have

$$AB \equiv AB.$$

We can state the above axiom briefly by saying that every segment can be laid off upon a given side of a given point of a given straight line in one and only one way.

.   .   .   .   .

Group V. Axiom of Continuity (Archimedes' Axiom).

V.1. Let A_1 be any point upon a straight line between the arbitrarily chosen points A and B. Take the points $A_2, A_3, A_4, \ldots$ so that A_1 lies between A and A_2; A_2 between A_1 and A_3; A_3 between A_2 and A_4, etc. Moreover, let the segments $AA_1, A_1A_2, A_2A_3, A_3A_4, \ldots$ be equal to one another. Then, among this series of points, there always exists a certain point A_n such that B lies between A and A_n.

The axiom of Archimedes is a linear axiom.

Remark. To the preceding five groups of axioms, we may add the following one, which, although not of a purely geometrical nature, merits particular attention from a theoretical point of view. It may be expressed in the following form:

Axiom of Completeness (Vollständigkeit): To the system of points, straight lines, and planes, it is impossible to add other elements in such a manner that the system thus generalized shall form a new geometry obeying all five groups of axioms. In other words, the elements of geometry form a system that is not susceptible to extension, if we regard the five groups of axioms as valid.

Questions of this type have been asked for centuries. The most famous of them is said to have been posed already in the time of Euclid and refers to the axiom that is known as the *parallel postulate*. In Euclid's *Elements*, the axiom is stated as follows:

> If the inner angles on one side of a straight line intersecting two other straight lines sum up to less than two right angles, the two straight lines intersect if sufficiently extended to this side.

Today the axiom is formulated in plane geometry as

> For any straight line *g* and a point *P* not on *g*, there is exactly one straight line through *P* which is parallel to *g* in the sense that it has no point in common with *g*.

As one may see from Figure I.4.1, when this axiom is included with the others it is tantamount to the statement that the sum of the three interior angles of a triangle is two right angles (180°). Now our question again:

> Can we deduce the parallel postulate from the other axioms of Euclidean geometry (as set up by Hilbert, for example) by bare logic alone?

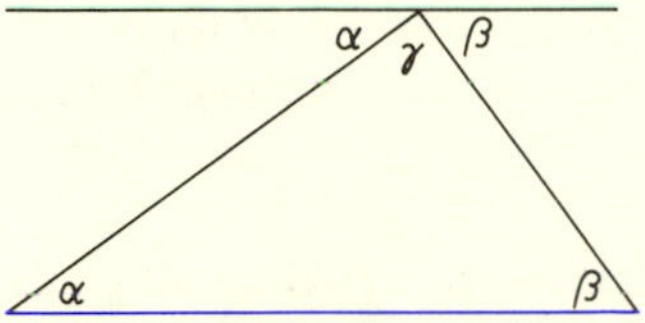

Figure I.4.1

The renowned Jesuit Gerolamo Saccheri (1667–1733) still believed that he had given such a deduction in his paper "Euclides ab omni naevo vindicatus" (1733). His proof that the angle sum of a triangle is ≤ 180° was correct within his system, but his proof that the angle sum is ≥ 180° was not. Today we know that the answer to the above question is *no*. How may one arrive at such an unprovability theorem? By setting up a theory that is rigorously based on Euclidean geometry and is therefore as rigorously based as Euclidean geometry, but that violates the parallel postulate even though it satisfies all the other axioms. Such a theory is called a *non-Euclidean geometry*.

The first established non-Euclidean geometry is that of Lobacevski [1829]. This work was followed soon after by Bolyai [1832]. For the intellectuals of the nineteenth century, it was almost impossible to accept non-Euclidean geometry as a purely logical counterexample. Whatever was called "geometry" had to comply with our spatial imaginations. From this perspective Immanuel Kant (1724–1804) had declared Euclidean geometry to be the only admissible theory. The question as to the validity of non-Euclidean geometry became quite controversial. Carl Friedrich Gauss (1777–1855) learned Russian in order to read Lobacevski in the original version. On 6 March 1832 he wrote to his friend Wolfgang Bolyai (1775–1856), who had sent him the non-Euclidean theory (Bolyai

[1832]) of his son János (1793–1856). Gauss pointed out that he himself had made the same investigations 30 years earlier without making them public because, as he put it in a letter to Bessel dated 27 January 1819 (in a different context), he "feared the shoutings of the Boeotians." This communication shocked the Bolyais. The Hungarian writer Laszlo Nemeth (b. 1901) wrote a drama on this story which was successfully performed in Budapest (Nemeth [1961]).

Kant's declaration mentioned above was made in his ponderous discussion of the following question:

> Naive question: Which geometry is valid in real space?
>
> Meta-question: Is this a question posed to nature or one posed to our way of judging reality?
>
> Meta-meta-question: How should such questions be decided? In what sense are they admissible, compulsory, or inevitable?

During the course of the nineteenth century, this type of reasoning led mathematicians to regard non-Euclidean geometry as being more than just a bare logical maneuver.

Non-Euclidean geometry became one of a variety of mathematical models for the interpretation of the facts of empirical geometry. One of the decisive steps in this direction was taken by Bernhard Riemann (1826–1866) in his famous Habilitationsvortrag "Über die Hypothesen, welche der Geometrie zugrundeliegen" (Riemann [1854]). Gauss (d. 1855), who had been instrumental in the choice of this theme, was deeply impressed. More than half a century later, "Riemannian geometry" provided one of the mathematical tools of Einstein's theory of general relativity (Einstein [1915, 1916]). For a modern account see Sachs-Wu [1977].

The classical non-Euclidean geometries occur in the curricula of our universities even today—for example, see Perron [1962] or Nöbeling [1975]. These geometries are divided into two types called *hyperbolic* geometries and *elliptic* geometries. In a hyperbolic geometry the angle sum of a triangle is less than 180° and more than one line can be drawn through a given point parallel to a given straight line. In an elliptic geometry the angle sum of a triangle is more than 180° and there are no parallels at all.

We may obtain an elliptic geometry, for example, by taking a sphere and redefining its great circles as our "lines." Here there are no parallels at all, and in this geometry it is easy to construct triangles with three right angles. The angle sum of such a triangle is 270° > 180°. This spherical geometry is not new at all to seafarers and astronomers, and the only unconventional aspect of the theory is the renaming of great circles as lines. If we confine ourselves to small subregions of the sphere, the geometry approximates Euclidean geometry.

We may obtain a two-dimensional hyperbolic geometry by restricting ourselves to the interior of a circle and taking for our "lines" the circles orthogonal to the boundary. See Figure I.4.2. In this geometry one may see that there is exactly one line through any two given points. However, there are infinitely many lines through a given point parallel to a line that does not pass through that point. Alternatively, we could take straight secants instead of circles orthogonal to the boundary. A hyperbolic geometry may also be obtained in a half-plane by taking circles orthogonal to the boundary as the "lines."

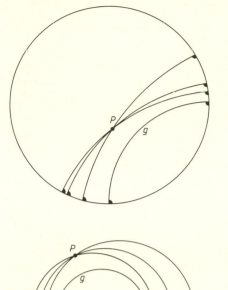

Figure I.4.2

4.3   Analytic Geometry

René Descartes (1596–1650) was not only one of the most influential philoso-phers of all times, he also enriched mathematics with a fundamental idea, the so-called analytic geometry (Descartes [1637]). In this approach, points in the plane are repre-sented by ordered pairs (x, y) of numbers via a coordinate system. Geometric figures them turn out to be the solutions to sets of equations or systems of equations, and so algebra is able to lend a helping hand to geometry. Figure I.4.3 displays some figures and their equations. In this table we have used the example of the hyperbola to show how the same geometrical object may be represented in algebraically different ways in differ-ent coordinate systems. For example, using the change of coordinates $x = \xi + \eta$, $y = \xi - \eta$, we may transform the equation $xy = 1$ into the equation $(\xi + \eta)(\xi - \eta) = \xi^2 - \eta^2 = 1$.

Such transformations are routine in analytic geometry and are constantly being used to reduce a given equation to its simplest possible form. Outside of mathematics, how-ever, the role of these transformations may go far beyond a search for a convenient form of an equation that makes it, for example, practicable from a computational point of view. For instance, such transformations may be used to write out the laws of nature in their most revealing and ''natural'' form, and they may have an unexpected bearing on questions of religion or Weltanschauung. An example of such a question is whether the sun revolves around the earth or vice versa. This is not a mathematical question: tell us how you like it, and we shall calculate it all for you. We shall even calculate it for you

Figure and coordinate system	Equation
Line 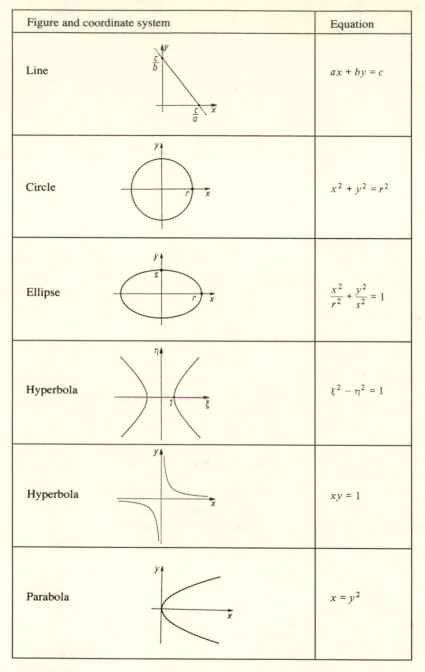	$ax + by = c$
Circle	$x^2 + y^2 = r^2$
Ellipse	$\dfrac{x^2}{r^2} + \dfrac{y^2}{s^2} = 1$
Hyperbola	$\xi^2 - \eta^2 = 1$
Hyperbola	$xy = 1$
Parabola	$x = y^2$

Figure I.4.3

both ways if you so order (and you pay us). Physicists prefer the heliocentric model because it makes the laws of planetary motion more simple and perspicuous. However, astronomers are constantly making use of transformations to and from geocentric coordinates as well. In spite of the mathematical equivalence of these two ideas, we know only too well how much bitter controversy they have generated on the religious plane.

By transforming equations of the type

$$ax^2 + by^2 + cxy + dx + ey = g$$

(where $a, b, \ldots, g$ are given constants) into the simple forms that are shown in Figure I.4.4, we may obtain the traditional *classification of quadratic curves*. Note that the idea of an isomorphism plays a role here. Figure I.4.4 displays the essential part of this classification. The conics (ellipse, parabola, and hyperbola) were already studied in antiquity by a number of mathematicians including Apollonios of Perga (ca. 262–ca. 190 B.C.), who was an important younger contemporary of Archimedes (ca. 287–212 B.C.).

A similar set of transformations may be used in three dimensions to *classify the quadric surfaces*, which are the surfaces that satisfy a quadratic equation in three variables x, y, and z. In this case we obtain a longer list of examples. We shall not list them all here, but we display three of them in Figure I.4.4, which shows these surfaces intersecting with one another orthogonally. The surfaces shown are an ellipsoid, a one-sheet hyperboloid, and a two-sheet hyperboloid.

Descartes's idea of analytic geometry had an enormous impact not only on geometry, but on all of mathematics. We sum up this impact in the following comments:

(1) In a way, the impossibility proofs of §2 could not have been formulated without analytic geometry.

(2) Analytic geometry bears on the foundations of geometry. It shows that geometry is logically as consistent as the theory of numbers.

(3) Finally, analytic geometry has opened the way to the geometry of dimensions higher than three. It is, in principle, no more difficult to handle four,

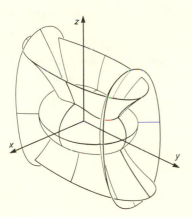

Figure I.4.4

five, or more coordinates than it is to handle two or three. For example, the equation

$$x^2 + y^2 + z^2 + t^2 = 1$$

is the equation of the unit sphere in four dimensions, and in five dimensions the equation of this sphere is

$$x^2 + y^2 + z^2 + t^2 + u^2 = 1.$$

In this way, analytic geometry allowed mathematics to break through a barrier that had been created by the limits of our imagination. While this breakthrough was important for mathematics, it was also important for the applications of mathematics into other fields, such as physics and economics.

4.4 Projective Geometry

The rigorous development of geometry that was produced by Hilbert [1899] allows us to investigate the "logical statics" of geometry. Such investigations had, of course, begun long before Hilbert. In fact, during much of the 19th century attempts were made to develop that part of geometry that can be deduced from the laws of combining and intersecting alone. During these attempts the concept of parallelism was felt to be a disturbing exceptional case. The geometers would have preferred a simple condition such as (in plane geometry)

two distinct straight lines intersect in exactly one point,

which would make it unnecessary to distinguish between the cases of parallel lines and intersecting lines. They felt the condition that two lines g and g' are parallel should be restated as the condition that g and g' intersect at a point at infinity. This feeling was motivated by the way parallel lines (such as railroad tracks) appear to converge at infinity in a picture (see Figure I.4.5).

Pictures like this support the notion that a line might stretch toward infinity to the right and then return from infinity from the left. In this way, lines were given a sort of circular structure.

How does one make rigorous mathematics out of such suggestions? Then answer is, by *defining*, *making*, or *adding* the "infinite points." For example, we can obtain an infinite point of intersection of parallel lines by *defining* this point to be the set of all lines

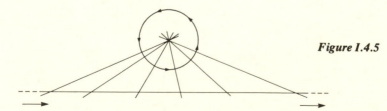

Figure I.4.5

that are parallel to a given line. The nonmathematician will, of course, object: Such a set of parallels *is* not a point. But the mathematician answers: That is unimportant. In mathematics it is irrelevant what things are. Only their logical roles are important. The preceding definition of an infinite point defines the concept in such a way that any two lines have exactly one point in common, regardless of whether the lines are parallel or not; therefore, this definition makes any distinctions between parallelism and nonparallelism obsolete. Note that in this approach to the problem, we do not discover the infinite points—we *make* them.

The way of thinking that mathematicians employ in order to formulate a definition like that of the preceding paragraph often puzzles the nonmathematician. Most puzzling of all is the way mathematicians have *arranged* for the truth of the assertion that any two distinct lines have exactly one point in common. By changing the definition, we could just as well have arranged for this assertion to be false. When we make the assertion true, we obtain a new theory called *projective geometry*.

In the plane, this theory can be deduced from the following brief system of just three simple axioms. (In higher-dimensional cases a more elaborate system of axioms is needed.)

> P I.   Any two different points are joined by exactly one line.
> P II. Any two distinct lines intersect in exactly one point.
> P III. There exist four points such that no three of them lie on the same line.

The patent terseness of this system of axioms does not indicate weakness, but rather a degree of abstraction that opens possibilities never thought of before. In order to obtain a system that satisfies these three axioms we do not need the huge Euclidean plane and the definition of an infinite point that was given in the preceding paragraphs. For example, the axioms can be satisfied by a space consisting of a bare seven points and seven "lines," each of which consists of three points—the so-called *projective seven-point-plane* (see Figure I.4.6). Note that in this figure, only the bold points count.

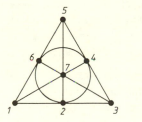

Figure I.4.6

The projective seven-point-plane can also be described by the following table:

> Points: 1,2,3,4,5,6,7
> Lines: 123,345,561,174,572,376,264

The ability of this model to satisfy the axioms P I–P III is in no way hampered by the fact that its three-point lines are not lines in the usual sense of the word. There are many other models, both finite and infinite. For example, the definition of an infinite point given previously also leads to a model satisfying the three axioms; and there are also projective geometries of dimensions higher than two. A fundamental result in projective geometry is *Desargues's theorem*, discovered in 1648 by Gérard Desargues (1593–1661). This theorem requires the use of three dimensions, as suggested in Figure I.4.7, and can be stated as follows:

> If $A_1B_1C_1$ and $A_2B_2C_2$ are any two triangles such that the three lines A_1A_2, B_1B_2, and C_1C_2 pass through a single point, then the points of intersection of the corresponding sides (suitably extended) of the two triangles must all lie in a single straight line g.

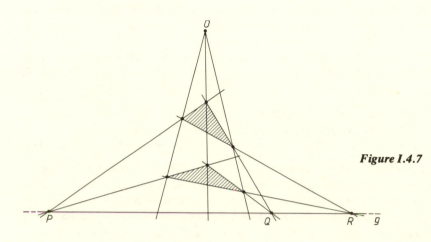

Figure I.4.7

In three dimensions, the line g is obtained by intersecting the two planes in which the triangles lie. It is interesting to note that in the plane, Desargues's theorem cannot be proved using the axioms P I–III alone: there are "non-Desarguian" planes. The smallest non-Desarguian plane has 91 points and is therefore very hard to visualize. We shall not describe any further details of the wealth of results that have been harvested in projective geometry to date. Instead, we shall be content with a few general statements:

(1) Axiom P I can be obtained from axiom P II, and vice versa, simply by interchanging the words "point" and "line." The terms "join" and "intersect" can be fused into a single symmetric term such as "a point and a line are incident." To axiom P III we may add a symmetric condition, which is provable from P I–III:

P IV. There are four lines, no three of which pass through one point.

By purely formal-logic arguments, we may now propagate the symmetry concept throughout all of (planar) projective geometry. What we see is that for every theorem of projective geometry, there is a "dual" theorem which may be obtained simply by interchanging the terms "point" and "line";

and this interchange also transforms the proof of the original theorem into the proof of the dual theorem. This symmetry principle is not a statement about the objects of projective geometry, but is instead a statement about the theory itself as it deals with these objects. The principle is not mathematics, but *meta-mathematics*, and is called the *duality principle of projective geometry*. Its complete formulation seems to be due to Joseph Diaz Gergonne (1771–1859, Gergonne [1825]), but the idea can be traced back to François Viète (1540–1603).

(2) The so-called *projective completion* of planes by infinite points can be achieved on the algebraic side of the theory. For this purpose we must work with full algebraic equipment. What we actually need is the concept of a field that will be introduced in Section 4 of Chapter II. It can be shown that not all projective planes lead to fields. Instead, the theory of projective planes has led to very interesting generalizations, known as "loops," of the concept of a field. Many statements about the "algebraic side" of a given projective plane can be read from certain geometric figures, known as "Schließungsfiguren." The figure associated with Desargues's theorem is one of them, and corresponds to the associative law. (See, for example, Pickert [1955] or Dembowski [1968].)

(3) Some projective planes are finite, like the seven-point-plane that we discussed previously. The theory of finite projective planes plays an important role in modern combinatorics, and in this way it leads to applications like the design of statistical experiments and the organization of telephone networks. See, for example, Jacobs [1983a] and Beth-Jungnickel-Lenz [1985].

4.5 The Group-Theoretical Systematics of Geometry: Felix Klein's Erlanger Programm (1872)

When Felix Klein (1849–1925) became a full professor in Erlangen in 1872 (at the age of 23 years) he published, according to an established Erlangen custom, what is known as a Programmschrift. His Programmschrift was titled "Vergleichende Betrachtungen über neuere geometrische Forschungen" (Comparative considerations about recent research in geometry) (Klein [1872]). This Erlanger Programm of Klein should not be confused with his inaugural address, which dealt with didactical questions (see Rowe [1985] and Jacobs-Utz [1984]).

The main ideas of Klein's Erlanger Programm had a profound influence on the systematization of geometry. These ideas can be formulated as follows:

Every subdiscipline of geometry is characterized by a particular transformation group; and every subdiscipline deals with precisely those properties of geometrical objects that are invariant under the transformations of its group.

For example, Euclidean geometry is characterized by the group of all rigid (in other words, distance-preserving) motions in the plane or in three-space. Since the angles in a triangle are determined by the lengths of its sides, statements about the size of angles belong in Euclidean geometry. The subdisciplines affine geometry and projective geometry are associated with other groups, and even topology (which we shall discuss in Chapter V) and set theory can be characterized in such a fashion. Today the ideas that Felix Klein set forth in his Erlanger Programm have a place in the thoughts of every mathematician.

§5   *Some More Views of Geometry*

Descartes's idea of an analytic geometry transformed geometry into a vast playground of ideas and methods of algebra and analysis. All of the techniques of analytic geometry relate to operations on numbers, or on members of other kinds of algebraic systems, and the last three centuries have witnessed a major transformation of geometry in this direction. For a comprehensive, modern presentation of the results of this transformation, see Dubrovin-Fomenko-Novikov [1985] or Berger [1987].

In this new world of geometry, the visual geometry in the style of Euclid looks merely like a marginal field. However, upon closer investigation, this marginal field presents itself as a vigorous mathematical discipline in its own right. Important researchers such as Wilhelm Blaschke (1885–1962) (see Strubecker [1986]) and Harold Scott MacDonald Coxeter (b. 1907) have left their imprint on this field. We end this chapter by listing some of the problem areas and results in "visual" geometry.

Packings in the plane and in three-space (for example, sphere packings of maximal density); see Fejes Tóth [1953] and Coxeter [1981].

Problems of tiling (see, for example, Grünbaum-Shephard [1983] and Voderberg's [1936, 1937] spiral tiling of the plane (Fig. I.5.1).

Danzer's four-nails theorem (Danzer [1986]): If you place an arbitrary number of discs on a table such that any two of the discs overlap, then four nails, suitably driven through the discs into the table, are sufficient to prevent all of the discs from sliding down onto the floor when the table it tilted.

Figure I.5.1

Minimal surfaces (such as soap bubbles or tent constructions). See Hilde-
brandt-Tromba [1985] and the bibliography given there.

A related topic is the so-called isoperimetric problem. Its classical special case may be
stated as follows: Among all simple closed planar curves of a given length, which one
includes the largest area? The answer to this question is that there is exactly one such
curve: the circle. An analogous role is played by spheres in three-space. Blaschke [1916]
is a classic on this subject. Generalizations of problems of this type are still the object of
active research; see Hadwiger [1957]. A related question asks: Is the honeycomb optimal? See Fejes Tóth [1964].

Chapter II · Elements of Algebra

THE NONMATHEMATICAL public often thinks of mathematicians as people who "can compute everything" or who "want to compute everything," but, in reality, a substantial part of mathematics is not concerned with computing at all. Of course, there are applied mathematicians and those who work in the world of the computer for whom calculations are a major focus of activity. A *pure mathematician*, however, usually looks upon computations as an occasional necessary evil. They are chores that must be performed from time to time (as seldom as possible) in order to achieve the main object of mathematics: *the proving or disproving of conjectures*. Typically, a mathematician may perform computations in order to construct examples that will allow him to test a conjecture, or to help him guess what a reasonable conjecture should say. Carl Friedrich Gauss obtained some of his most important ideas from the vast calculations that he performed (Maennchen [1930]). Computers might have freed him from the tedium of many of these boring calculations.

Apart from their interest in using calculations to test their conjectures and motivate new ones, mathematicians are also interested in many questions of principle that surround calculations. Questions like *What is possible and what is meaningful in computing?* or *Can a computer think?* have transformed the art of calculating into a spiritually demanding discipline, "computer science." Within this discipline even pure mathematicians discover now and then the enjoyment of calculating and of working with concrete numbers. But the soul of a pure mathematician will never be satisfied with a mere ability to calculate well. A mathematician is constantly searching for the questions of principle that underlie the calculations that stand before him, questions such as

> What does "calculating" mean?
> What does solving equations mean?
> What do we need in order to make a given type of calculation possible?
> How many steps and how much time does it take to perform a given type of calculation?

In this chapter we shall become acquainted with some of the basic ideas of algebra, and we shall gain some insight into the mathematical treatment of the first three of these four questions. In §1 we shall recall the four basic operations of arithmetic, $+$, $-$, $\cdot$, $\div$. We shall also explore the minimal requirements for these four operations to be meaningful—requirements that will lead us to the concept of a *field*. Besides the classical field $\mathbb{Q}$ of rational numbers we shall also see some other kinds of fields, including the finite fields GF(2) and GF(3). We shall also deduce a large number of rules of calculation that are valid in any field. For example, we shall see why the rule "minus times minus equals plus" has to hold in any field. This kind of rule is the sort of statement that everyone learns in school, but usually without any proof. In §2 we shall go one step further and define carefully what it means to take a square root of an element of a field. Then in §3 we shall investigate the problem of taking square roots in more detail. The number $\sqrt{2}$ cannot exist in $\mathbb{Q}$—the Pythagoreans knew that—but by so-called *quadratic field extension* we can take arbitrary square roots, even $\sqrt{-1} = i$. Finally, in §4 we shall deal with the problem of solving equations and systems of equations, and with the problem of calculating approximate square roots numerically.

As everywhere in this book, we shall explain the key considerations in considerable detail. Explanation of the notation and of many of the techniques of calculation and proof will be replaced by an array of typical examples.

The facts treated in this chapter are widely dispersed over the whole of mathematics. Driver [1984] covers many of them at a senior high school level. See also Engel [1984], where the algorithmic viewpoint—almost untouched in this book—is treated with particular emphasis. The most comprehensive treatise on the algorithmic approach is Knuth [1968]. Fields and field extensions are treated systematically in every university algebra textbook. Here Waerden [1970] is a classic. See also Waerden [1983, 1985].

§1   *The Four Basic Arithmetical Operations and the Concept of a Field*

The four basic arithmetical operations are addition, subtraction, multiplication, and division. For many kinds of everyday calculations, like accounting, balancing of books, and calculation of (simple) interest, these four operations are all that we need. For the actual performance of these operations we nowadays often rely on computers. An entire wealth of calculation techniques developed during the past centuries is incorporated in the algorithmic devices of computer science (see Knuth [1968], Engel [1984]).

For mathematicians, these operations of arithmetic motivate the following question of principle:

> What is the least that we need to assume in order to obtain a mathematical system that will allow us to carry out operations like the operations of arithmetic?

In attempting to answer this question we are led to the mathematical concept of a *field*. In this section we shall take a detailed look at the field concept. We begin with a few important observations about the operations of arithmetic.

First: subtraction is the "inverse" of addition and division is the "inverse" of multiplication. By this we mean that

$a - b$ is the element that must be added to b in order to yield a:

$$b + (a - b) = a$$

and

$a \div b$, also written as $\dfrac{a}{b}$, is the element that must be multiplied by b to yield a.

Therefore, if we want to have a mathematical system that supports the four operations of arithmetic, it is sufficient to have just an addition and a multiplication as long as we know that these two operations can be "inverted." More precisely, we need to know that for any two elements a and b there is an element d such that $b + d = a$. Such an element d is called the *difference* between a and b. We have to be a little more careful in saying what we mean by inverting multiplication. We cannot simply require that for any two elements a and b there is an element q (the *quotient* of a and b) such that $b \cdot q = a$. The trouble is that we may have $b = 0$, in which case $b \cdot q = 0$ whatever element q we choose. We must therefore avoid the case $b = 0$; and this is why we have the rule

"never divide by 0 (never write the expression $\dfrac{a}{0}$)."

This rule is one of the many rules one learns about in school without seeing why the rule has to be there. It might be argued that we should define

$$\frac{1}{0} = \infty,$$

but while such statements do play a role in some branches of mathematics, they are merely symbolic, or they are a substitute for a more complicated statement. We shall not deal with infinite numbers for the present.

Second: In order to ensure that addition and multiplication can be inverted, we need guarantee only the following two statements:

For every b there is an element called $-b$ such that $b + (-b) = 0$.
For every $b \neq 0$ there is an element called b^{-1} such that $b \cdot b^{-1} = 1$.

The elements $-b$ and b^{-1} are respectively known as the *additive inverse* and *multiplicative inverse* of b. Once the existence of these inverses has been assumed, we can define the difference $a - b$ of two elements a and b to be $a + (-b)$ and we can define the quotient $a \div b$ to be $a \cdot b^{-1}$. In fact, this leads to

$$b + (a - b) = b + (a + (-b)) = a + (b + (-b)) = a + 0 = a$$

and

$$b \cdot (a \div b) = b \cdot (a \cdot b^{-1}) = a \cdot (b \cdot b^{-1}) = a \cdot 1 = a.$$

We shall now put all these observations together. If we want a mathematical system that will allow us to carry out the four operations of arithmetic (apart from division by zero), then we need a set K that is endowed with two special elements 0 and 1 and two operations $+$ and $\cdot$ that provide us with the *sum* $a + b$ and the *product* $a \cdot b$ of any two elements a and b. Furthermore, we need to assume the arithmetical laws that are summarized in Table II.1.1.

We now add a few words of explanation to the laws as they appear in Table II.1.1. Note first that in the first three laws the elements a and b need to be introduced, or, as mathematicians say, *quantified*. For example, a precise statement of the commutative laws should read as follows:

For all elements a, b of the set K, we have

$$a + b = b + a \quad \text{and} \quad a \cdot b = b \cdot a.$$

These commutative laws say that any two elements a and b can be *commuted* (interchanged) without affecting their sum $a + b$ or their product $a \cdot b$. The associative laws tell us that when considering the expressions $(a + b) + c$ and $(a \cdot b) \cdot c$ we may move the parentheses and write $a + (b + c)$ and $a \cdot (b \cdot c)$ without affecting these numbers. We may therefore write these expressions without any parentheses at all, defining

$$a + b + c = (a + b) + c = a + (b + c)$$
$$a \cdot b \cdot c = (a \cdot b) \cdot c = a \cdot (b \cdot c).$$

Table II.1.1

Addition	Multiplication	Name of law
$a+b = b+a$	$a \cdot b = b \cdot a$	commutative law
$a+(b+c)=(a+b)+c$	$a \cdot (b \cdot c)=(a \cdot b) \cdot c$	associative law
$a+0 = a$	$a \cdot 1 = a$	identity elements for $+$ and $\cdot$
For every a there is an element $-a$ such that $a+(-a)=0$	For every $a \neq 0$ there is an element a^{-1} such that $a \cdot a^{-1} = 1$	existence of additive and multiplicative inverses
$a \cdot (b+c) = (a \cdot b)+(a \cdot c)$		distributive law

Combining the commutative and associative laws we can now see that if a, b, and c are any elements of K, then they can be added or multiplied in any order. For example,

$$a + b + c = a + c + b = c + a + b.$$

This obviously generalizes to more than three elements.

   We should also add a remark concerning the existence of additive and multiplicative inverses. A more precise statement of these laws would be as follows (see above):

> For every element a in the set K there is at least one element b such that $a + b = 0$.
>
> For every element a in the set K, if $a = 0$, then there is at least one element b such that $a \cdot b = 1$.

As we shall see soon, if a is an element of K then there cannot be more than one element b such that $a + b = 0$. The unique element b such that $a + b = 0$ can therefore be called *the* additive inverse of a and can be given the name $-a$, "minus a." Similarly, if $a = 0$ then there is a unique element b such that $a \cdot b = 1$. This unique b is called *the* multiplicative inverse of a and is given the name a^{-1}.

We remark finally that the distributive law can be written a little more simply if we adopt the usual convention that in the absence of parentheses, multiplication is done before addition. Thus $a \cdot b + c$ means $(a \cdot b) + c$. We also agree to omit the multiplication symbol $\cdot$ whenever we can. With these conventions, the distributive law says that for any elements a, b, and c of K we have $a(b + c) = ab + ac$.

We are now ready to give the definition of a *field*.

A field consists of a set K that is endowed with two special elements 0 and 1 and two operations $+$ and $\cdot$ that satisfy the rules of Table II.1.1.

The most familiar example of a field is, of course, the field $\mathbb{Q}$ of rational numbers. As you may know, a rational number is a number that may be written as a quotient, $\dfrac{m}{n}$ or m/n, where m and n are integers (and $n \neq 0$). From the identities

$$\frac{m}{n} + \frac{m'}{n'} = \frac{mn' + m'n}{nn'}$$

$$\frac{m}{n} \cdot \frac{m'}{n'} = \frac{mm'}{nn'}$$

we see that the sum and product of two rational numbers are always rational and, as we know, the rules of Table II.1.1 all hold. Historically, the evolution of the number system was as follows:

(a) The natural numbers were invented for counting and adding.
(b) The number zero and the negative integers were invented in order to allow all subtractions to be performed.
(c) Fractions (rational numbers) were invented in order to allow all divisions to be performed (except division by zero, of course).

We should understand, however, that in the number system there is more going on than just the four operations of arithmetic upon which the definition of a field is based. For example, our abstract definition of a field makes no mention of the order relation that exists in the number system. We should therefore not be too surprised to see some examples of fields that look quite different from the field $\mathbb{Q}$. Actually, from some points of view, $\mathbb{Q}$ is quite a complicated field. The simplest possible field is a field in which $0 = 1$. If K is such a field, then given any element a in K we have $a = a \cdot 1 = a \cdot 0 = 0$, and so the only element in this field K is 0. In other words, $K = \{0\}$. In deducing this fact, we have used the identity $a \cdot 0 = 0$. We shall show soon that this identity holds in every field.

Apart from the trivial field that has $0 = 1$, the next simplest field is the field that contains just the two (distinct) elements 0 and 1. This field is usually called GF(2), "Galois field two," after Evariste Galois (1811–1832). Addition and multiplication in GF(2) are defined as follows:

Explicitly

Addition Table

$0 + 0 = 0,$

+	0	1
0	0	1
1	1	0

$0 + 1 = 1 + 0 = 1$

$1 + 1 = 0$

Explicitly

Multiplication Table

$0 \cdot 0 = 0$

·	0	1
0	0	0
1	0	1

$0 \cdot 1 = 1 \cdot 0 = 0$

$1 \cdot 1 = 1$

Most of these definitions come as no surprise, being immediate consequences of the properties of 0 and 1, or of the identity $a \cdot 0 = 0$ that holds in every field. The only one of these that calls for some explanations is the definition $1 + 1 = 0$. To see that this equation is as inevitable as the others, we use the fact that 1 must have an additive inverse: for some element b in GF(2) we have $1 + b = 0$. But $b = 0$ doesn't work because $1 + 0 = 1 \neq 0$, and therefore $b = 1$.

We see therefore that the only possible way to make a field GF(2) out of the two-element set $\{0,1\}$ is to define addition and multiplication according to the preceding tables. We have not actually verified that this way of defining the operations gives us a field, but it is easy to check that the laws of Table II.1.1 hold, and we leave this task to the reader. For example, the commutative laws are visible from the symmetry of the addition and multiplication tables with respect to the diagonal $\diagdown$.

To a nonmathematician it may seem a little strange to have $1 + 1 = 0$. Since we are accustomed to $1 + 1 = 2$, we seem to be saying that $2 = 0$. If these statements run counter to your intuition, please remember that the elements of a field do not have to be numbers. The element 1 in a field does not have to be the number 1. It is merely the multiplicative identity of the field. In the field GF(2), the element $1 + 1$ is not the number 2. In fact, the element $1 + 1$ is equal to the element 0. Only when we look at a field of numbers like $\mathbb{Q}$ can we say that the element 1 is the same as the number 1 and that the element $1 + 1$ is the same as the number 2.

A nonmathematician might also question the value of looking at a field like GF(2). He might ask, "Even if there is a field like this, what is it good for?" We can give a partial answer to this question by looking at the addition and multiplication of odd and even integers. In the system of integers we have

EVEN + EVEN = EVEN
ODD + EVEN = EVEN + ODD = ODD
ODD + ODD = EVEN,
EVEN · EVEN = EVEN
ODD · EVEN = EVEN · ODD = EVEN
ODD · ODD = ODD

If we replace the word ODD by the symbol O and the word EVEN by the symbol E, then the preceding equations can be arranged into the following tables:

+	E	O
E	E	O
O	O	E

·	E	O
E	E	E
O	E	O

If we now replace the symbols E and O by 0 and 1, respectively, then these two tables are identical to the tables we used to define the field GF(2). Actually, this way of looking at GF(2) led to the discovery of this field around the year 1800. The field GF(2) also has applications to computer science because of the binary character of electronics: for example, an electrical circuit can exist in either of two states, closed or open. When a circuit is closed the current flows, and when the circuit is open the current cannot flow.

By taking a closer look at even and odd integers, we can see how to extend the idea and construct some other finite fields. An integer is even when it can be put in the form $2n$, where n is an integer, and is odd if it has the form $2n + 1$. We shall now consider multiples of 3. If k is a given integer, then there are three possibilities:

(a) k is a multiple of 3. In this case $k = 3n$ for some integer n.
(b) When k is divided by 3, the remainder is 1.
    In this case $k = 3n + 1$ for some integer n.
(c) When k is divided by 3, the remainder is 2.
    In this case $k = 3n + 2$ for some integer n.

If we now associate the integers of these three types with the numbers 0, 1, and 2, respectively, then the usual addition and multiplication leads us to the tables

+	0	1	2
0	0	1	2
1	1	2	0
2	2	0	1

·	0	1	2
0	0	0	0
1	0	1	2
2	0	2	1

For example, the equation $2 \cdot 2 = 1$ can be interpreted to say that the product of two integers of "type 2" is an integer of "type 1." In fact,

$$(3m + 2)(3n + 2) = 3(3mn + 2m + 2n + 1) + 1.$$

It is easy to verify that these definitions of $+$ and $\cdot$ in the set $\{0,1,2\}$ satisfy all the rules of Table II.1.1, and therefore provide us with a field. This field is usually called GF(3).

If we consider multiples of 4 and apply the same techniques, we find ourselves with a definition of $+$ and $\cdot$ in the set $\{0,1,2,3\}$, but this time the results are disappointing. We obtain the tables

+	0	1	2	3
0	0	1	2	3
1	1	2	3	0
2	2	3	0	1
3	3	0	1	2

·	0	1	2	3
0	0	0	0	0
1	0	1	2	3
2	0	2	0	2
3	0	3	2	1

but although the addition table does what it should, the multiplication does not. Note, for example, that the element 2 has no multiplicative inverse. So this method of constructing a field GF(4) with four elements does not work. There is, however, another method of defining operations $+$ and $\cdot$ in the set $\{0,1,2,3\}$ that does give us a field. This method yields the following addition and multiplication tables:

+	0	1	2	3
0	0	1	2	3
1	1	0	3	2
2	2	3	0	1
3	3	2	1	0

·	0	1	2	3
0	0	0	0	0
1	0	1	2	3
2	0	2	3	1
3	0	3	1	2

It is easy to verify that these operations satisfy the rules of Table II.1.1. If we apply the previous method to multiples of 5, then we obtain a definition of $+$ and $\cdot$ that turns the set $\{0,1,2,3,4\}$ into a field called GF(5). It can be shown, however, that there is no field GF(6) with six elements. Actually, it can be shown that if n is any natural number, then there is a field GF(n) with n elements if and only if n has the form p^m, where p is some prime and m is a natural number. $2 = 2^1, 3 = 3^1, 4 = 2^2, 5 = 5^1, 7 = 7^1, 8 = 2^3$, $9 = 3^2$ are such prime powers; $6 = 2 \cdot 3$ and $10 = 2 \cdot 5$ are not. In the event that n has this form, it can also be shown that for practical purposes, there is only one field with n elements. By this we mean that once we have selected a field GF(n) with n elements, any other field K with n elements can be converted into GF(n) with a mere change of notation. Two fields that are related in this way are said to be *isomorphic*.

Now that we have acquainted ourselves with the notion of a field and seen some examples, we shall begin the task of deducing some facts that are valid in *every* field. Many of these facts are just those rules of arithmetic that one learns in school without proof, but we shall now show that these facts follow logically from the rules in Table II.1.1.

1. *In any given field, the element 0 is the only additive identity.*
   Suppose that K is a given field and that $0'$ is any additive identity. Since 0 is an additive identity we have

   $$0' + 0 = 0',$$

and since $0'$ is an additive identity we have

$$0 + 0' = 0.$$

But from the commutative law we see that

$$0 + 0' = 0' + 0,$$

and so we deduce that $0 = 0'$.

2. *In any given field, the element 1 is the only multiplicative identity.*
   Suppose that K is a given field and that $1'$ is any multiplicative identity. Arguing as we did before, we can see that $1' = 1$.

3. *An element of a field can have only one additive inverse.*
   Suppose that a is an element of a field K and that b and c are both additive inverses of a. We need to show that $b = c$. Now

$$b = 0 + b = (a + c) + b$$
$$= a + (c + b) = a + (b + c)$$
$$= (a + b) + c = 0 + c = c.$$

It might be a good exercise to take a good look at each step in this chain of reasoning, and to ask yourself which of the rules of Table II.1.1 we are using in each case.

4. *A nonzero element of a field can have only one multiplicative inverse.*

5. *If a is an element of a field, we have $-(-a) = a$.*
   Since

$$-a + a = a + (-a) = 0,$$

we see that a is an additive inverse of $-a$. But $-(-a)$ is the only additive inverse of $-a$ by proposition 3, and so we conclude that $-(-a) = a$.
   As an exercise, you should state and prove a similar fact about multiplicative inverses.

6. *Multiplication by 0 always yields 0.*
   Suppose that a is an element of a field K. We need to show that $0 \cdot a = 0$. First we observe that

$$0 \cdot a = (0 + 0) \cdot a = 0 \cdot a + 0 \cdot a.$$

Now, using the existence-of-inverses rule in Table II.1.1, we choose an element b of K such that $0 \cdot a + b = 0$. For any element c of K we now obtain

$$c = c + 0 = c + 0 \cdot a + b = c + (0 \cdot a + 0 \cdot a) + b$$
$$= c + 0 \cdot a + (0 \cdot a + b)$$
$$= c + 0 \cdot a + 0 = c + 0 \cdot a$$

Thus $0 \cdot a$ is an additive identity and $0 \cdot a = 0$ follows from 1.

7. *Given any elements a and b of a field, we have*

$$(-a) \cdot b = a \cdot (-b) = -(a \cdot b).$$

Suppose that a and b are elements of a field K. To see that $(-a) \cdot b$ is the additive inverse of $a \cdot b$, we need to see that $(-a) \cdot b + a \cdot b = 0$. Using the distributive law and proposition 6 we obtain

$$(-a) \cdot b + a \cdot b = ((-a) + a) \cdot b = 0 \cdot b = 0.$$

A similar argument shows that $a \cdot (-b) = -(a \cdot b)$. As a special case, we note

$$(-1) \cdot a = -a.$$

8. *Given any elements a and b of a field we have*

$$(-a) \cdot (-b) = a \cdot b.$$

Using proposition 7 twice and then proposition 5 we obtain

$$(-a) \cdot (-b) = -(a \cdot (-b)) = -(-(a \cdot b)) = a \cdot b.$$

An interesting special case of proposition 8 occurs with $a = b = 1$. In this case we obtain $(-1) \cdot (-1) = 1$.

9. *In any field, two times two is four.*
The element "two" in a field K is the element $2 = 1 + 1$, and the element "four" is the element $4 = 1 + 1 + 1 + 1$. We see that

$$\begin{aligned} 2 \cdot 2 &= (1 + 1) \cdot (1 + 1) = 1 \cdot (1 + 1) + 1 \cdot (1 + 1) \\ &= (1 + 1) + (1 + 1) = 1 + 1 + 1 + 1 = 4. \end{aligned}$$

We see therefore that the statement "two times two is four" is a mathematical theorem and not merely a convention (as, for example, Goethe believed). We should mention, however, that in order to define the elements 2 and 4 of a field, we are making implicit use of the natural numbers two and four as well.

10. *The product of two nonzero elements of a field with 0 ≠ 1 is always nonzero.*
Suppose that a and b are nonzero elements of a field K. Since

$$aba^{-1}b^{-1} = (aa^{-1}) \cdot (bb^{-1}) = 1 \cdot 1 = 1 \neq 0,$$

we deduce at once from proposition 6 that it is impossible to have $ab = 0$. Finally we take a look at some common formulas:

11. *Given any elements a and b of a field, we have*

$$(a + b)^2 = a^2 + 2ab + b^2$$
$$(a - b)^2 = a^2 - 2ab + b^2$$
$$(a + b)(a - b) = a^2 - b^2.$$

All of these follow simply from the distributive and commutative laws. We shall prove the first one and leave the other two as exercises.

$$\begin{aligned} (a + b)^2 &= (a + b) \cdot (a + b) \\ &= (a + b)a + (a + b)b \end{aligned}$$

$$= aa + ba + ab + bb$$
$$= a^2 + ab + ab + b^2$$
$$= a^2 + 1 \cdot ab + 1 \cdot ab + b^2$$
$$= a^2 + (1 + 1) \cdot ab + b^2$$
$$= a^2 + 2ab + b^2.$$

The preceding eleven propositions show that many of the familiar rules of algebra follow logically from the conditions in Table II.1.1. A nonmathematician might ask whether it was really worth the trouble of proving these propositions. Shouldn't we perhaps have avoided all these proofs by adding the statements of these propositions to Table II.1.1? However, had we done so, then every time we wanted to show that a given mathematical system is a field, we would have had to check every one of the propositions. We would have found ourselves working very hard proving the propositions again and again. Therefore, what we have done in this section saves us a great deal of trouble, for we now know that once the conditions of Table II.1.1 are satisfied, all of the propositions follow automatically.

§2   *Square Roots*

In this section we go a step beyond the four basic arithmetical operations $+$, $-$, $\cdot$, $\div$ and focus our attention upon the square and square roots of elements of a given field. As you know, the square of an element a of a given field is simply the element $a \cdot a$, and so squaring is nothing but a special case of multiplication. On the other hand, the extraction of square roots is something entirely new. A square root of an element a of a given field means an element x such that $x^2 = a$, and, in this sense, square rooting can be looked upon as being the inverse operation of squaring. Another way of looking at square roots is that they are solutions of quadratic equations.

Given an element a of a field K, a *square root* of a is a solution of the equation $x^2 = a$ or, equivalently, a solution of the equation $x^2 - a = 0$. The idea of a square root presents us with two fundamental difficulties:

(1) Not every element of a field has a square root. For example, the element -1 has no square root in the field $\mathbb{Q}$. The reason for this is that in $\mathbb{Q}$ it makes sense to talk about positive and negative numbers. For every element x of $\mathbb{Q}$ we know that x^2 is either positive or zero, and therefore, since the number -1 is negative, there cannot exist an element x in $\mathbb{Q}$ such that $x^2 = -1$.

(2) If a nonzero element a of a field has a square root, then it has two square roots. In fact, if b is a square root of a, then since $(-b)^2 = (-b) \cdot (-b) = b^2 = a$, we see that $-b$ is also a square root of a. However, if $b^2 = a$ then the numbers b and $-b$ are the only square roots of a. There are no others because the equation $x^2 - a = 0$ can be written as $x^2 - b^2 = 0$, which yields $(x - b)(x + b) = 0$. From Proposition 10 of the previous section we see that this equation can be satisfied only when $x - b = 0$ or $x + b = 0$, and therefore b and $-b$ are the only values of x that can satisfy

this equation. In the same way we can see that the only square root of 0 is 0.

Because of these two difficulties we have to be a little careful in our use of the symbol $\sqrt{a}$. If an element a of a field K has a square root, then we can let the symbol $\sqrt{a}$ stand for one of the square roots of a. So if x is any element of the field satisfying $x^2 = a$, then either $x = \sqrt{a}$ or $x = -\sqrt{a}$; and we write this statement as $x = \pm\sqrt{a}$.

If a and b are elements of a field K and both a and b have square roots which we write as $\sqrt{a}$ and $\sqrt{b}$, then since

$$(\sqrt{a} \cdot \sqrt{b})^2 = (\sqrt{a})^2 \, (\sqrt{b})^2 = ab,$$

we see that $\sqrt{a} \, \sqrt{b}$ is a square root of ab. Informally, we may write this condition in the form

$$\sqrt{ab} = \sqrt{a} \, \sqrt{b},$$

and in the same way we can see that

$$\sqrt{\frac{a}{b}} = \sqrt{a} \, / \, \sqrt{b}.$$

In a field like $\mathbb{Q}$ where it makes sense to talk about positive and negative numbers, the symbol $+\sqrt{a}$ is sometimes used to denote the nonnegative square root of an element a. In other words, if $b^2 = a$, then $+\sqrt{b}$ is whichever of the two numbers b and $-b$ is nonnegative.

Square roots occur frequently in the applications of mathematics. We display a few examples.

(1) Suppose we have a rectangular sheet of paper of length a and width c as depicted in Figure II.2.1. The problem is to find the ratio a/c that has the property that if we cut the sheet into two rectangles with length c and width $a/2$ as shown, then the ratio of length to width of the new rectangles is the same as that of the original one. What we require is that

Verbally	Graphically	Formulas
If we halve the sheet parallel to its shorter edge, we obtain two sheets with the same ratio of edge lengths as the original sheet	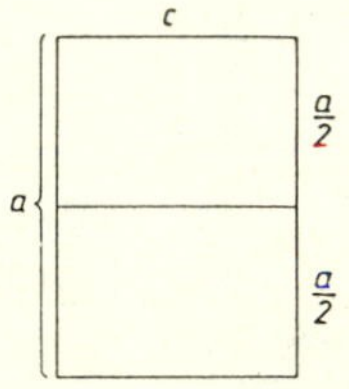	$a : c = c : \dfrac{a}{2}$ i.e., $\dfrac{a}{c} = 2\dfrac{c}{a}$ i.e., $a^2 = 2c^2$ i.e., $\left(\dfrac{a}{c}\right)^2 = 2$ i.e., $\dfrac{a}{c} = \sqrt{2} = 1.4142\ldots$

Figure II.2.1

$$a : c = c : \frac{a}{2} \, ;$$

in other words,

$$\frac{a}{c} = \frac{2c}{a} \, ,$$

which yields

$$\left(\frac{a}{c}\right)^2 = 2.$$

The required ratio $a : c$ is therefore $\sqrt{2} = 1.4142 \ldots$ Note that the ratio of length to width of the standard 8.5-by-11-inch sheets that are used in the United States is only about 1.294. So American sheets are "too square."

(2) The "golden section," $1 : 1.6180 \ldots$, is also the solution of a problem that concerns the ratio of the sides of a rectangle. In this case we start with a rectangle with length a and width c, and we cut off a square with side c leaving a new rectangle with a length of c and a width of $a - c$. See Figure II.2.2. The problem is to find the ratio c/a such that the ratio of length to width of the new rectangle is the same as that of the old one. What we require is that

Verbally	Graphically	Formulas
Subtracting the shorter side from the longer one, we obtain a new rectangle with the same ratio of side lengths	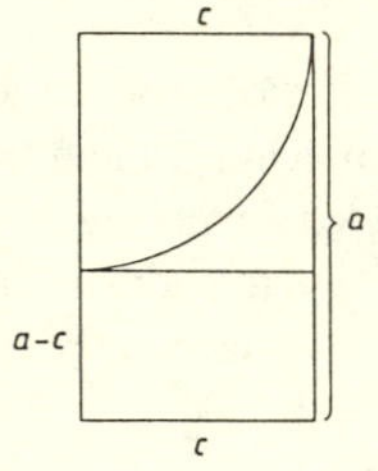	$a : c = c : (a - c)$ i.e., $\dfrac{a}{c} = \dfrac{c}{a - c}$ i.e., $a(a - c) = c^2$ i.e., $\dfrac{a}{c}\left(\dfrac{a}{c} - 1\right) = 1$

Figure II.2.2

$$a : c = c : (a - c),$$

i.e.,

$$\frac{a}{c} = \frac{c}{a - c},$$

$$a(a - c) = c^2,$$

$$\frac{a}{c}\left(\frac{a}{c} - 1\right) = 1.$$

If we put $x = a/c$, we obtain

$x(x - 1) = 1,$
$x^2 - x - 1 = 0.$

When this quadratic equation is solved by completing the square, we obtain

$$\left(x - \frac{1}{2}\right)^2 = 1 + \frac{1}{4} = \frac{5}{4},$$

$$x - \frac{1}{2} = \sqrt{\frac{5}{4}} = \frac{\sqrt{5}}{2}$$

i.e.,

$$x = \frac{1 + \sqrt{5}}{2} = 1.6180. \ldots$$

This rectangle is a little more oblong than the one we saw in Example (1).

(3) Science is full of square roots that arise out of a use of the Pythagorean theorem. A famous one of these is the "Einstein root,"

$$\sqrt{1 - \frac{v^2}{c^2}},$$

which plays an important role in the theory of special relativity.

We mention finally that approximate values like $\sqrt{2} = 1.4142 \ldots$ require us to step beyond the domain of pure algebra. We will say more about this topic in §4.

§3   *Taking Square Roots: Questions of Principle*

In §1 we dealt with the question

What sort of mathematical system do we need that will allow us to carry out the four arithmetical operations $+$, $-$, $\cdot$, and $\div$ (except for division by zero)?

We answered this question by introducing the notion of a field, and we looked at some examples of fields, like the field $\mathbb{Q}$ of rational numbers and the fields of the form GF(n). A natural question to ask at this stage is what do we need in a field in order to allow us to take the square root of every element of the field. Some fields have this property, like GF(2), while others do not. For example, look at the field GF(3) = $\{0,1,2\}$. Since $0^2 = 0$, $1^2 = 1$, and $2^2 = 1$, there is no element of GF(3) whose square is 2, and therefore the element 2 of GF(3) has no square root.

The field that interests us most if the field $\mathbb{Q}$ of rational numbers, and in this field square roots are quite hard to come by. For a start, the field $\mathbb{Q}$ contains an order relation $<$ that allows us to speak about positive and negative elements, and, as we have already pointed out, negative elements can never have square roots. But in addition to this, there are many positive elements of $\mathbb{Q}$ that do not have square roots. For example, the number 2 has no square root in $\mathbb{Q}$. This fact was probably discovered by the Pythagoreans around 400 B.C.

3.1   The Nonexistence of $\sqrt{2}$ in $\mathbb{Q}$

We shall show that there is no rational number a such that $a^2 = 2$. In order to prove this fact, we shall assume that such a number a exists, and then we shall show that our assumption leads to a contradiction. Assume then that a is a rational number and that $a^2 = 2$, and choose two integers m and n such that

(1) $a = \dfrac{m}{n}.$

In the event that two integers m and n have a common factor, this common factor can be "canceled" in the fraction m/n. We shall assume that this has been done. In other words, we assume that a has been written as a fraction m/n whose numerator and denominator have no common factor. From the fact that $an = m$, we have $a^2 n^2 = m^2$, and therefore

(2) $2n^2 = m^2.$

Since the left side of this equation is an even number, we see that the number m^2 is also even. Therefore the number m itself must be even because if m were odd, then m^2, being the product of two odd numbers, would be odd. Using the fact that m is even we choose an integer p such that $m = 2p$. Squaring we obtain $m^2 = 4p^2$, and substituting this expression for m^2 into (2) we obtain

$2n^2 = 4p^2,$

and on dividing both sides by 2 we obtain

(3) $n^2 = 2p^2.$

We now repeat the argument that we used before to deduce that the number n is also even, and we choose an integer q such that $n = 2q$. We now have

$m = 2p$ and $n = 2q$

in spite of our earlier assumption that the numbers m and n have no common factor. This is the required contradiction.

The kind of proof that we have just used is called "proof by contradiction" or "indirect proof." This kind of argument was used frequently by Plato (429–348 B.C.), who displayed a liking for indirect "dialectic" proofs. In using this method, Plato was really following in the footsteps of his teacher Socrates (470–399 B.C.), who was fond of refuting his opponents by letting them appear to be right at first, and then demonstrating how they had entangled themselves in a web of contradictions. This kind of debating strategy brought Socrates dangerously close to his enemies, the Sophists, and also created new enemies for him.

The proof we have given of the nonexistence of $\sqrt{2}$ in $\mathbb{Q}$ can easily be extended to the square root of any prime number. Thus we have the following theorem:

In the field $\mathbb{Q}$ of rational numbers, no prime number has a square root.

You probably know that there is an infinite number of prime numbers. Therefore, in realizing that the method of proof we used to establish the nonexistence of $\sqrt{2}$ also applies to other primes, we are at once led to an infinite number of conclusions. This kind of reasoning is not a proof in the ordinary sense; instead, it is reasoning about the possibility of proving and works on a higher level than the original proof. We call this higher level "meta-level." Reasoning of this type is called "meta-mathematical" reasoning; in the form displayed here it is a daily tool of the mathematician.

The proof we have given of the nonexistence of $\sqrt{2}$ in $\mathbb{Q}$ was purely algebraic (or number-theoretical), but for the Greeks, geometry was the whole of mathematics. They had a geometric proof of the nonexistence of $\sqrt{2}$ in $\mathbb{Q}$. Suppose that we have a square with a side of one unit and a diagonal of a units (see Figure II.3.1). By the Pythagorean theorem we have $a^2 = 1^2 + 1^2 = 2$, and therefore $a = \sqrt{2}$. To obtain a contradiction, let us assume that a is a rational number m/n. Then we can decompose each side of the square into n segments of equal length and the diagonal of the square into m segments of the same length. By changing the scale in our figure we may assume that each of these segments has a length of one inch. To obtain the desired contradiction we shall show that a figure like this is impossible. We shall now draw in a few additional lines (see Figure II.3.2).

It is easy to prove that the two angles denoted by α are equal, and that the two segments marked b are also equal. It may also be seen that the upper left corner of Figure II.3.2 is a square of side $c = a - b$. We leave these proofs to the reader. We now make the following conclusions:

(1) The length b is the difference between the length of the diagonal and the length of the side of the big square. Therefore b is a whole number of inches.

(2) Since $c = a - b$, the length c is also a whole number of inches.

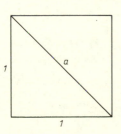

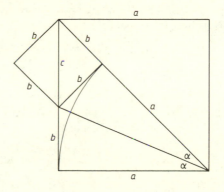

Figure II.3.1							**Figure II.3.2**

Thus the figure "square with diagonal" has been reproduced on a smaller scale, but with its side and diagonal still whole numbers of inches. However, this argument can be repeated again and again, until both of these lengths are less than one inch, and we arrive at the desired contradiction.

Historically, the Pythagoreans are said to have made their fundamental discovery not with $\sqrt{2}$ but with $\sqrt{5}$. The number $\sqrt{5}$ occurs in the golden section and in the pentagon, and was, from the viewpoint of Weltanschauung, more interesting and exciting to the Pythagoreans. See Fritz [1945].

3.2   Quadratic Field Extension

As we have seen, not every element of the field $\mathbb{Q}$ has a square root in $\mathbb{Q}$. In fact, we have encountered two separate barriers against the taking of square roots of elements of $\mathbb{Q}$. One of these is the fact that negative members of $\mathbb{Q}$ can never have square roots in $\mathbb{Q}$, but, on a more sophisticated level, we have also found positive elements such as the prime numbers $2,3,5,7, \ldots$ that have no square roots in $\mathbb{Q}$.

We shall now describe a method that will allow us to overcome these barriers. We shall show that if a is an element of an arbitrary field K, then even if a does not have a square root in K, it will have a square root in some other field F that is bigger than K. To motivate the theorem that follows let us consider the number $\sqrt{2}$ again. We have seen that this number does not exist in $\mathbb{Q}$, but, as you know, $\sqrt{2}$ does exist in the field $\mathbb{R}$ of all *real numbers*. Now let us define F to be the set of all those real numbers that can be written in the form $r + s\sqrt{2}$, where r and s are rational numbers. It is worth noticing that two elements $r + s\sqrt{2}$ and $r' + s'\sqrt{2}$ of F can be equal only if $r = r'$ and $s = s'$, for suppose that

$$r + s\sqrt{2} = r' + s'\sqrt{2}.$$

Then unless $s = s'$ we obtain

$$\sqrt{2} = \frac{r - r'}{s' - s},$$

which is impossible since $\sqrt{2}$ is not rational. Therefore $s = s'$, and it follows at once that $r = r'$. Now from the identities

$$(r + s\sqrt{2}) + (r' + s'\sqrt{2}) = (r + r') + (s + s')\sqrt{2}$$

$$(r + s\sqrt{2})(r' + s'\sqrt{2}) = (rr' + 2ss') + (rs' + r's)\sqrt{2}$$

we see that addition and multiplication can be carried out in the set F. We note also that the numbers $0 = 0 + 0 \cdot \sqrt{2}$ and $1 = 1 + 0 \cdot \sqrt{2}$ are elements of F. Finally we observe that if $r + s\sqrt{2}$ is any nonzero element of F, then

$$\frac{1}{r+s\sqrt{2}} = \frac{1(r-s\sqrt{2})}{(r+s\sqrt{2})(r-s\sqrt{2})} = \frac{r}{r^2 - 2s^2} + \left(\frac{-s}{r^2 - 2s^2}\right)\sqrt{2}.$$

The only thing we have to worry about when we look at the preceding identity is the possibility that $r^2 - 2s^2 = 0$. But this cannot happen, because it would imply that $\sqrt{2}$ $= \pm r/s$, which is an element of $\mathbb{Q}$. We have therefore seen that F is a field, and it is clear that F is the smallest field of numbers that can contain the number $\sqrt{2}$.

You should understand that what we have just said is merely a motivation for a theorem in which we shall prove the existence of a field that contains a square root of 2. In this motivation we have worked with the symbol $\sqrt{2}$ in much the same way that one works with mathematical symbols in high school, studying its behavior without knowing that it exists. As a matter of fact, this is how mathematicians did their calculations for centuries. We shall now show that the sort of argument given in this motivation can be written precisely, and that it can be used to give us a square root of an element of an arbitrary field.

Theorem. Suppose that a is an element of a field K and that a is not the square of any member of K. Then there is a field $\bar{K}$ including K such that $\bar{K}$ contains an element b satisfying $b^2 = a$.

Sketch of Proof. We define $\bar{K}$ to be the set of all ordered pairs (r,s) of elements of K. Since the order of the elements r and s is important in the expression (r,s), we note that two ordered pairs (r,s) and (r',s') are the same if and only if $r = r'$ and $s = s'$. In order to understand this definition of the set $\bar{K}$, keep in mind the motivation that you have just read. We would like an element of $\bar{K}$ to have the form $r + s\sqrt{a}$, but since we do not actually have an object $\sqrt{a}$ in our hands, we write the ordered pair (r,s) instead. We now define addition and multiplication in $\bar{K}$ as follows:

$$(r,s) + (r',s') = (r + r', s + s')$$
$$(r,s)(r',s') = (rr' + ass', rs' + r's).$$

A glance at the previous motivation will tell you that these definitions of $+$ and $\cdot$ are just what they should be. From the identities

$$(r,s) + (0,0) = (r + 0, s + 0) = (r,s)$$
$$(r,s) \cdot (1,0) = (r \cdot 1 + a \cdot s \cdot 0, s \cdot 1 + r \cdot 0) = (r,s)$$

we see that the elements $(0,0)$ and $(1,0)$ are neutral elements for the operations $+$ and $\cdot$ respectively. Given any element (r,s) of $\bar{K}$, we deduce from the identity

$$(r,s) + (-r, -s) = (0,0)$$

that $(-r, -s)$ is the additive inverse of (r,s). We shall now show that if $(r,s) \neq (0,0)$, then the element

$$\left(\frac{r}{r^2 - as^2}, \frac{-s}{r^2 - as^2} \right)$$

is the multiplicative inverse of (r,s). We need to check first that the denominator $r^2 - as^2$ is not zero. But if $r^2 - as^2 = 0$, then since the elements r and s are not both zero, neither of them can be zero, and we obtain $a = r^2/s^2 = (r/s)^2$, contradicting the fact that the element a does not have a square root in K. Therefore $r^2 - as^2 \neq 0$. Now to verify that (r,s) has a multiplicative inverse we observe that

$$\left(\frac{r}{r^2 - as^2}, \frac{-s}{r^2 - as^2} \right) \cdot (r, s)$$

$$= \left(\frac{r^2}{r^2 - as^2} - \frac{as^2}{r^2 - as^2}, \frac{-rs}{r^2 - as^2} + \frac{rs}{r^2 - as^2} \right) = (1,0).$$

It is now easy to verify that the system $\bar{K}$ satisfies all the laws in Table II.1.1—in other words, that $\bar{K}$ is a field. We leave this task to the reader.

We now look at an important subset K^* of F. We define K^* to be the set of all members of $\bar{K}$ having the form $(r,0)$. We note that both the additive identity $(0,0)$ and the multiplicative identity $(1,0)$ are elements of K^*, and that for any elements $(r,0)$ and $(r',0)$ of K^* we have

$$(r,0) + (r',0) = (r + r',0)$$
$$(r,0) \cdot (r',0) = (rr',0).$$

From these identities we see that K^* is a carbon copy of the field K. We can think of an element $(r,0)$ of the field K^* as being the element r of K together with the "ornament" $(\quad,0)$. If we are willing to overlook this ornament, then we can think of K and K^* as being the same field. In this sense we can say that K is contained in the field $\bar{K}$. Perhaps we should say that K is contained in $\bar{K}$ "up to an ornament," or, as the mathematicians would say, "up to isomorphy."

Finally we come to the most important step of all. We shall show that the element a (or $(a,0)$, if we write it with its ornaments) has a square root in the field $\bar{K}$. Remember that when we defined $\bar{K}$ to be the set of all ordered pairs (r,s) where r and s are elements of K, what we had in mind was that the ordered pair (r,s) would represent the expression $r + s \sqrt{a}$ that we could not write officially. Therefore $\sqrt{a} = 0 + 1 \cdot \sqrt{a}$ should be represented by the ordered pair $(0,1)$. We shall show, in fact, that

$$(0,1) \cdot (0,1) = (a,0).$$

Note that

$$(0,1) \cdot (0,1) = (0 \cdot 0 + a \cdot 1 \cdot 1, 0 \cdot 1 + 0 \cdot 1) = (a,0),$$

as required.

The method of the preceding section could even be used to add a square root of -1 to the field $\mathbb{Q}$, and therefore the theorem provides us with a logical basis for the introduction of the imaginary number $i = \sqrt{-1}$ and the field of complex numbers. The method of this theorem can also be extended. As we have seen it, the theorem tells us that a field can always be extended in order to provide us with solutions of equations of the form $x^2 - a = 0$.

University students of algebra learn how to apply this method, suitably generalized, to general equations of the form

$$x^n + a_{n-1}x^{n-1} + \ldots + a_1 x + a_0 = 0.$$

Fields that have been obtained for this purpose are known as *algebraic field extensions* of the given field.

§4   *Solution of Equations and Systems of Equations*

A wide variety of problems can be reduced to the problem of solving equations such as $ax = c$ or $x^2 = a$. In this section we discuss some equations that can be solved by the methods of Sections 1, 2, and 3.

1. The problem of solving an equation such as $ax = c$ or $x^2 = a$ is the problem of finding a number x (or, more generally, an element x of a given field) that satisfies the given equation. Such an element is called a *solution* of the said equation. It may happen that several solutions exist, and in this case we have to find *all* of the solutions. In other words, we have to find the *solution set* of the equation.

For example, if we wish to solve the equation $ax = c$, then there are a number of possibilities:

> In the event that $a \neq 0$, the equation has exactly one solution, $x = c/a$.
> In the event that $a = 0$ and $c \neq 0$, the equation has no solution.
> In the event that $a = 0$ and $c = 0$, every element of the given field is a solution of the equation.

Alternatively, we might consider the equation $x^2 = 4$. This equation has exactly two solutions, $x = 2$ or $x = -2$.

2. One method of solving an equation is to rewrite it in several equivalent forms until one of these has the form $x = c$. At this stage the solution of the equation has been found. In order for this method of solution to make sense we need to know what we mean by "equivalent forms" of an equation. We say that two equations are equivalent when they have the same solution set.

To illustrate this idea we shall look at some examples. In the first of these we solve the equation

(1) $3x + 12 = 0.$

We start off by observing that if we add -12 to both sides of equation (1), then we obtain an equivalent equation

(2) $3x + 12 - 12 = 0 - 12,$

which we can write as

(3) $3x = -12.$

The preceding argument explains what we are really doing when we say that we are solving equation (1) by "taking the term 12 to the right side and changing its sign." We are now ready for our next step. We divide each side of equation (3) by 3 and obtain the equivalent equation

(4) $\frac{1}{3}(3x) = \frac{1}{3}(-12),$

which we can rewrite in the form

(5) $x = -4.$

The only solution of equation (5) is -4, and since equations (5) and (1) have the same solution set, we conclude that the only solution of equation (1) is -4.

In our second example we solve an equation that requires us to take square roots. We solve the equation

$$2x^2 - 4x + 6 = 0.$$

We begin by writing the following sequence of equations that are equivalent to one another:

$$x^2 - 2x + 3 = 0.$$
$$x^2 - 2x = -3.$$

(6) $x^2 - 2x + 1 = -3 + 1 = -2$

(7) $(x - 1)^2 = -2.$

The step that takes us from equation (6) to equation (7) is called *completing the square*, and follows from the fact that $(x - 1)^2 = x^2 - 2x + 1$. Now equation (7) is equivalent to the condition that

$$x - 1 = \pm \sqrt{-2},$$

in other words,

$$x = 1 \pm \sqrt{-2}.$$

Using the methods that were developed in Section 3.2 we can write

$$\sqrt{-2} = \sqrt{(-1) \cdot 2} = i \sqrt{2},$$

and we can therefore say that the solution set of the given equation consists of the two numbers $1 + i\sqrt{2}$ and $1 - i\sqrt{2}$. As you may know, this method of completing the square can be generalized to the general quadratic equation

$$ax^2 + bx + c = 0$$

(where $a \neq 0$), and it yields the well-known "quadratic formula"

$$x = \frac{-b + \sqrt{b^2 - 4ac}}{2a}.$$

Furthermore, if the so-called *discriminant* $D = b^2 - 4ac$ is negative, then this solution involves the "imaginary" number i.

Similar but much more complicated formulas are known for the more general equations

$$ax^3 + bx^2 + cx + d = 0$$

and

$$ax^4 + bx^3 + cx^2 + dx + e = 0,$$

even though these formulas may involve cube roots of *complex* numbers even when all of the solutions of the equation are *real*. These formulas are usually named after Girolamo Cardano of Padova (1501–1576), even though they had been discovered earlier by Niccolo Fontana of Brescia (1499–1557), who was commonly known as Tartaglia (stammerer) because of a speech impediment he had as a result of an injury he received as a young child. In 1826, Niels Henrik Abel (1802–1829) showed that for polynomial equations of degree 5 or higher, there is no general solution formula that uses only the four operations of arithmetic and radicals. This led subsequently to entirely new developments in algebra.

3. In addition to the problems that require us to solve a single equation for one unknown, there are many important problems that require us to solve a system of several equations for several unknowns. Some of these problems have been around for centuries. For example, on page 104 of a book that was written in 1525 by Adam Riese (1492–1559), the following problem is posed:

> At a party 21 people, men and women, have had a total of 81 drinks. Each man has had 5 drinks and each woman has had 3. How many men and how many women are there at the party?

Riese wrote the following solution:

$$\text{21 people} \qquad \begin{array}{c} \text{man 5} \\ \text{woman 3} \end{array} \qquad 81$$

Subtracting 3 from 5, we obtain 2, the divisor. Now multiply 3 by 21, which gives 63, and subtract this from 81. We obtain 18. Divide by 2. We see that there are 9 men, and by subtraction we see that there are 12 women.

Today we would solve this problem as follows:

> Let x be the number of men.
> Let y be the number of women.
> Then we see that

$$x + y = 21 \text{ and therefore } 3x + 3y = 63$$
$$5x + 3y = 81$$

Subtracting the left side of the first equation from the left side of the second and the right side of the first equation from the right side of the second, we obtain

$$5x - 3x = 83 - 63$$
$$2x = 18$$
$$x = 9, \text{ and } y = 21 - 9 = 12.$$

It is easy to substitute these values of x and y into the equations and thus verify that they work. This algebraic method may be a little clumsy, but it has the advantage of working in a wide variety of problems. For the simple problem suggested by Adam Riese, it would have been good enough to reason as follows:

Suppose first that all 21 people drink as if they are women. In this event the total number of drinks consumed is $21 \cdot 3 = 63$. We see that $81 - 63 = 18$ drinks are left over, and therefore $18/2 = 9$ people can now take an additional 2 drinks each. Therefore there are 9 men in the party.

Perhaps this is what Adam Riese had in mind. The same trick applies to the following well-known problem:

There are 10 animals in a barnyard, some of which are rabbits and some of which are chickens. If the total number of their legs is 26, how many chickens are there?

In the Chinese mathematics book *Chiu Chang Suan Shu (Nine Books of Mathematical method*, published in German in 1968) that dates back to the early Han period (202 B.C.– 9 A.D.) we find the following problem:

A man sold 2 oxen and 5 sheep, bought 13 pigs, and came home with 1000 pieces of money. A second man sold 3 oxen and 3 pigs, bought 9 sheep, and broke even. A third man sold 6 sheep and 8 pigs, bought 5 oxen, and had to pay in 600 pieces of money. What does each of the three animals cost? Answer: An ox costs 1200, a sheep costs 500, and a pig costs 300.

The book also provides the following explanation:

To solve the problem, follow the Rule of Fang Ch'êng. For the first man, the 2 oxen and the 5 sheep are positive, the 13 pigs are negative, and the remaining money is positive. For the second man, the 3 oxen and 3 pigs are positive and the 9 sheep are negative. For the third man, the 5 oxen are negative, the 6 sheep and 8 pigs are positive, and the additional money is negative. Apply the plus-minus rule to this information.

Today we would solve this problem as follows:

Let x be the price of an ox.
Let y be the price of a sheep.
Let z be the price of a pig.

Then

$$2x + 5y - 13z = 1000$$
$$3x - 9y + 3z = 0$$
$$-5x + 6y + 8z = -600.$$

By applying the usual high school methods of eliminating the unknowns one at a time, we can obtain

$$x = 1200 \qquad y = 500 \qquad z = 300.$$

The reader is invited to obtain these values and to verify that they work by substituting them into the equations.

The *Chiu Chang Suan Shu* [1968] has exercises of this type that lead to as many as five equations in five unknowns. The methods given in that book indicate that the early Han Chinese were, in principle, able to solve arbitrarily complicated problems of this sort.

The techniques for solving systems of several equations in several unknowns are the same as those we have described for single equations in one unknown. The solution may be obtained by rewriting the equations in several equivalent forms until finally we arrive at a system of the form $x = a$, $y = b$, . . . , $z = c$, where the numbers a, b, . . . , c are known elements of the given field and do not involve the variables x, y, . . . , z.

The following three "rules of thumb" help us to classify systems of equations in several unknowns. However, as you read them you should be aware that they are not as precise as we have stated them here, and they should therefore be applied with caution.

(a) If a system contains more equations than unknowns, then it often happens that there is no solution at all. Such a system may be "over-determined," and require too much of its unknowns. As an example of an over-determined system, look at the equations

$$x + 2y = 3$$
$$3x - 5y = 1$$
$$4x - 3y = 2$$

(b) If a system contains fewer equations than it has unknowns, then it often has infinitely many different solutions. If $m < n$, then a system of m equations in n unknowns will often leave $n - m$ "degrees of freedom" for its unknowns.

(c) When a system has the same number of equations as unknowns, then it usually has just one solution. The problems that we solved in this section were all of this type.

4. *Systems of Linear Equations.* All of the examples that we saw from Riese [1525] and *Chiu Chang Suan Shu* [1968] in the previous section led to systems of *linear* equations. Roughly speaking, an equation in variables x, y, . . . , z is said to be linear if it involves these variables to the first power only. This means that none of its terms involve x^2, x^3, y, yz, $1/y$, etc. When an equation contains more than three unknowns we find it more convenient to list them as x_1, x_2, x_3, . . . , x_n rather than to call them x, y, . . . , z. Using this more precise notation we can now say exactly what we mean by a linear equation. An equation in variables x_1, x_2, x_3, . . . , x_n is said to be *linear* if it has the form

$$a_1x_1 + a_2x_2 + a_3x_3 . . . + a_nx_n = c,$$

where a_1, a_2, a_3, . . . , a_n and c are known members of the given field.

There is a very well developed and effective theory for the solution of systems of linear equations. This theory finds the solutions when they exist and tells us when they do not. The theory also gives us precise statements to replace the rough rules of thumb that we stated at the end of the last section. We begin our description of this theory by describing an important technique that was perfected by Carl Friedrich Gauss (1777–1855) and that has come to be called *Gaussian elimination*.

(a) *Gaussian Elimination*. A system of m linear equations in n unknowns can be written in the form

$$a_{11}x_1 + a_{12}x_2 + a_{13}x_3 \ldots + a_{1n}x_n = c_1$$
$$a_{21}x_1 + a_{22}x_2 + a_{23}x_3 \ldots + a_{2n}x_n = c_2$$
(8) $\ldots$
$$a_{m1}x_1 + a_{m2}x_2 + a_{m3}x_3 \ldots + a_{mn}x_n = c_m.$$

Gaussian elimination converts the system (8) into a "triangular" form

$$b_{11}x_1 \qquad\qquad\qquad\qquad = R_1$$
$$b_{21}x_1 + b_{22}x_2 \qquad\qquad\qquad = R_2$$
(9) $\ldots$
$$b_{m1}x_1 + b_{m2}x_2 + b_{m3}x_3 \ldots + b_{mm}x_m = R_m,$$

where the terms $R_1, R_2, \ldots, R_n$ either are known, or (in the event that $m < n$) contain the unknowns $x_{m+1}, \ldots, x_n$. Once the system has been rewritten in the form (9), it may turn out that the numbers $b_{11}, b_{22}, \ldots, b_{mm}$ are all nonzero. In this case we go ahead as follows: We use the first equation to obtain

$$x_1 = \frac{R_1}{b_{11}}.$$

Now we substitute this value of x_1 into the second equation and take it over to the right side. In this way the second equation gives us

$$x_2 = \frac{R_2}{b_{22}} - \frac{R_1 b_{21}}{b_{11} b_{22}}.$$

We now substitute these values of x_1 and x_2 into the third equation, take them to the right side, and solve for x_3. We continue in this fashion until x_m has been calculated. We see now that there are two possibilities. If $m = n$ then the quantities $R_1, \ldots, R_n$ are all known, and this method gives exactly one solution of the variables $x_1, x_2, \ldots, x_n$. In the event that $m < n$, then the variables $x_{m+1}, \ldots, x_n$ can be given any values we like, and for any choice of these, the other variables $x_1, x_2, \ldots, x_m$ can be calculated. This is what we mean when we say that a system of equations has $n - m$ degrees of freedom.

Although this method of Gauss is highly efficient, it may still be improved. See Strassen [1969].

(b) *Representation of a System of Equations in Matrix Form*. A system of equations of the form (8)

$$a_{11}x_1 + a_{12}x_1 + a_{13}x_3 \ldots + a_{1n}x_n = c_1$$
$$a_{21}x_1 + a_{22}x_2 + a_{23}x_3 \ldots + a_{2n}x_n = c_2$$
(8) $\ldots$
$$a_{m1}x_1 + a_{m2}x_2 + a_{n3}x_3 \ldots + a_{mn}x_n = c_m$$

can be written more efficiently if we combine the coefficients on its left side into a *matrix*

$$A = \begin{bmatrix} a_{11} & a_{12} & \cdots & a_{1n} \\ a_{21} & a_{22} & \cdots & a_{2n} \\ & \cdot & & \\ & \cdot & & \\ & \cdot & & \\ a_{m1} & a_{m2} & \cdots & a_{mn} \end{bmatrix},$$

the numbers $c_1, \ldots, c_m$ on the right side into a so-called *column vector*

$$C = \begin{bmatrix} c_1 \\ c_2 \\ \cdot \\ \cdot \\ \cdot \\ c_m \end{bmatrix},$$

and the unknowns $x_1, x_2, \ldots, x_n$ into another column vector x. The method we have described for solving the system (8) can now be written in the language of matrices, and the system (8) itself can be written in matrix form as

(8a). $Ax = C.$

This shorthand version of (8) looks just like a single linear equation

$ax = c,$

which would be solved by writing

$x = a^{-1}c.$

The method of solving (8a) can now be reduced to the problem of finding a matrix A^{-1} that behaves like a multiplicative inverse for A. The solution of (8a) would then be

$x = A^{-1}C.$

The details of this method belong to the mathematical field known as *linear algebra*. The methods of linear algebra are not difficult, but the subject is too big for us to describe more fully here. We refer the reader to any good book on linear algebra for more information on this topic; see, for example, Greub [1967], Kowalsky [1970], or Koecher [1983]. An important feature of the solution of linear equations is that we need only the four operations $+$, $-$, $\cdot$, and $\div$ of arithmetic, and we can therefore work in any given field. For example, we could work in $\mathbb{Q}$. There is no need to resort to quadratic, algebraic, or any other type of field extension.

(c) Geometric Interpretation. Descartes's idea of representing points in the plane by pairs of numbers, and points in space by three numbers, allows us to interpret the solution sets of systems of equations geometrically. See Section 4.3 of Chapter I for more on this work of Descartes. The word "linear," which we have used only in a purely algebraic sense up till now, also has a geometric interpretation—it refers to *lines* (from the Latin word *lineae*). In fact, a single linear equation of the form $ax - by = c$ is satisfied by a

straight line of points (x,y) in the plane (as long as a and b are not both zero). See Figure II.4.1 for examples.

In a similar way, the set of points that satisfy a linear equation of the form $ax + by + cz = d$ is a plane in space, and the set of points that satisfy two equations of this form is the intersection of two planes. This interpretation of linear algebra in terms of classical geometry fails as soon as we deal with more than three unknowns, but mathematicians still use geometric language and use the geometric interpretation that exists in one, two, or three dimensions to motivate their work on a more abstract level. Although there may be some who feel that geometry in an n-dimensional space is not geometry in the purest sense of the word, the geometric approach has led to some formidable successes, which have led mathematicians to work quite freely in n-dimensional spaces, where n is any natural number, and even in infinite-dimensional spaces.

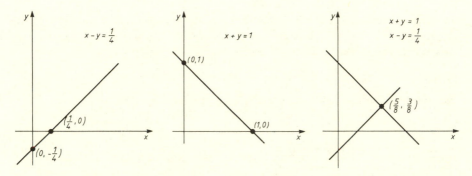

Figure II.4.1

We should mention that the solutions of nonlinear equations can also be interpreted geometrically in the sense of Descartes. This interpretation leads to a very active branch of mathematics that is called *algebraic geometry* (see, for example, Brieskorn-Knörrer [1981] and Shafarevic [1974]).

5. *Numerical Approximation of Square Roots*. The problem of finding the square root of a number a may be visualized as the problem of finding where the parabola $y = x^2$ intersects with the horizontal line $y = a$. See Figure II.4.2. The symmetry of this figure about the y-axis is a result of the algebraic identity $(-x)^2 = x^2$. We recall that if $\sqrt{a}$ exists, then $-\sqrt{a}$ is also a square root of a.

Now let us assume for the moment that $a = 2$. Since the number 2 has no square root in $\mathbb{Q}$, there is no point (x,y) with rational coordinates at which the parabola $y = x^2$ and the line $y = 2$ intersect. We can, however, look for points that *approximate* the points of intersection. In other words, we can look for rational numbers x for which the number x^2 is as close to a as we want it to be. See Table II.4.3.

This table displays some approximations to the number $\sqrt{2}$. Much more interesting than a table of this sort is a systematic procedure, an *algorithm*, for finding successive

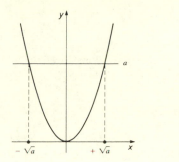

Figure II.4.2

Table II.4.3

$x \nearrow \sqrt{2}$	x^2	y^2	$y \searrow \sqrt{2}$
1	1	4	2
1.4	1.96	2.25	1.5
1.41	1.9881	2.0164	1.42
1.414	1.999396	2.002225	1.415
1.4142	1.999962	2.000245	1.4143

approximations to the square root of a given number a. One such algorithm, said to have been discovered by Heron of Alexandria around 130 A.D., is motivated by the equivalence of the equations

$$x^2 = a,$$

$$x = \frac{a}{x},$$

and

$$x = \frac{1}{2}\left(x + \frac{a}{x}\right).$$

This algorithm tells us that for any approximation x to $\sqrt{a}$, the number

$$\frac{1}{2}\left(x + \frac{a}{x}\right)$$

will be a better approximation. So if we start off with a number x_1 as our first approximation, then

$$x_2 = \frac{1}{2}\left(x_1 + \frac{a}{x_1}\right)$$

will be a better second approximation,

$$x_3 = \frac{1}{2}\left(x_2 + \frac{a}{x_2}\right)$$

will be a still better third approximation, and in general, if we have calculated the number x_n for any given natural number n, then our next approximation x_{n+1} is given by

$$x_{n+1} = \frac{1}{2}\left(x_n + \frac{a}{x_n}\right).$$

Before we can use an algorithm of this type we need to prove, of course, that it works. We need to prove that the numbers x_n come closer and closer to $\sqrt{a}$ as n increases, and we need to know just how rapidly x_n approaches $\sqrt{a}$. Justification of this algorithm rests on the following two observations:

(a) If for any given value of n, the number x_n is "too large" (meaning that $x_n^2 > a$), then since $(a/x_n)^2 < a$, the number a/x_n is too small. Similarly, if x_n *is too small, then* a/x_n is too large.

(b) Since x_n and a/x_n always lie on different sides of a, their arithmetical mean

$$x_{n+1} = \frac{1}{2}\left(x_n + \frac{a}{x_n}\right)$$

is closer to $\sqrt{a}$ than the worse of the two numbers x_n and a/x_n.

Chapter III · The Principle of Mathematical Induction

THE PRINCIPLE of mathematical induction is one of the important tools that are used in the construction of mathematical proofs. When we prove a theorem using the principle of mathematical induction, we often say that we are proving the theorem "by induction" or that we are "proceeding inductively from n to $n + 1$." However, you should understand that the word "induction," which we use to describe this principle, is really a misnomer. When the word "induction" is used in its proper sense, it refers to a type of reasoning that is commonly invoked in everyday life and in the sciences. For example, a scientist may tell us that if we drop a ball, then the force of gravity will cause it to fall to the ground. What he is really telling us is that *in the past* this has always happened and that be *believes* that what we have observed up till now will continue into the future. This type of "inductive" reasoning is strictly forbidden in mathematics. The principle of mathematical induction that we shall study in this chapter is really an example of *deductive* reasoning, the only type of reasoning that is allowed in mathematics. We use the word "induction" to describe the principle, only because it seems to have some of the flavor of the truly inductive reasoning that is used in science. Superficially, the principle seems to tell us that if something keeps on happening, then it will always happen.

In precise terms, the principle of mathematical induction is a statement about the system of natural numbers $1, 2, \ldots$ and can be stated as follows:

> Suppose that for every natural number n we are given a statement A_n. Suppose that the statement A_1 is known to be true and that for every natural number n, if the statement A_n happens to be true, then the statement A_{n+1} must also be true. Then the statement A_n must be true for every natural number n.

It is difficult to say just who first formulated this induction principle, but the ancient Greeks were certainly aware of the principle in one form or another. For more on this question see Freudenthal [1953] who declares Blaise Pascal (1623–1662) the author of the principle. See also Rabinovitch [1970].

We begin in §1 with three classical problems. In each case we shall first use an intuitive method to prove the theorem, and then we shall use the principle of induction to prove the theorem again. §2 deals with some questions of principle related to mathematical induction. In §3 we acquaint the reader with a classical field of applications of the induction principle: the basic laws of combinatorics. In §4 we give a proof by induction of a result known as the *marriage theorem*, which is one of the fundamental results in the modern theory of combinatorics (see also Section 2 of Chapter IV). In §5 we present an inductive proof of the *binomial theorem*, and in §6 we prove some fundamental properties of the system of natural numbers, including the *pigeonhole principle* and Dedekind's definition of finite sets. We also take a brief look at an interesting result known as *Ramsey's theorem*. Finally, in §7 we deal with *recursion*, which is the use of mathematical induction to define mathematical entities. We present three applications: the *Fibonacci numbers*, the *Thue-Morse sequence* (with a glimpse of some theorems by van der Waerden and Szemerédi on arithmetical progressions), and *Georg Cantor's diagonal argument*.

We recommend that nonmathematicians read Dedekind [1888] and Landau [1930].

§1   *Three Summations*

In this section we begin by proving three well-known and widely used statements about the summation of some special sequences of numbers. Having proved the three statements, we shall prove them again by induction. Then, having illustrated the induction principle by means of these three examples, we shall disucss it in detail in the next section.

1.1   *Arithmetic Progressions*

A finite sequence of numbers is said to be an *arithmetic progression* if it has the form

$$a, a+d, a+2d, \ldots, a+(n-2)d, a+(n-1)d,$$

where a and d are any given numbers and n is a natural number. The number a is the first member of the sequence. The number n is the number of members in the sequence, and we call it the *length* of the sequence. The number d is called the *common difference* of the sequence. We often refer to the members of a sequence as its *terms*. The problem that we shall be considering is finding the sum of all the terms of an arithmetic progression. In other words, our problem is to evaluate the sum

$$a + (a + d) + \ldots + (a + (n - 1)d).$$

We shall start with an arithmetic progression whose first term and common difference are 1. This is the progression

$$1, 2, 3, \ldots, n.$$

There is an anecdote about Carl Friedrich Gauss (1777–1855) that allegedly refers to this arithmetic progression in the case $n = 100$:

> According to the tradition in the schools at that time, when a mathematics problem was given to a class, the pupil who finished first placed his slate board down in the middle of a large table, and then the next to finish put his slate down on top of it. One day, when young Carl Gauss was a pupil in Mr. Büttner's arithmetic class, Mr. Büttner posed the problem of adding an arithmetic progression. He had hardly finished describing this task when Gauss threw his

slate board on the table saying, in low Braunschweigian dialect, "Ligget se" ("there she lies"). While the other pupils continued to work on this problem, Mr. Büttner, conscious of his dignity, walked up and down the room, and occasionally threw a contemptuous and caustic glance at the smallest of his pupils, who had finished the task too quickly. At last the other slates began to come in; and when the slates where turned over, Mr. Büttner found that Gauss' solution was correct even though many of the others were wrong (and were corrected with a slapping). (Waltershausen [1856])

We can surmise that little Gauss had reasoned in the following way: We want to know the value of

$$S = 1 + 2 + \ldots + 100.$$

Reversing the order of the terms, we can also write this number as

$$S = 100 + 99 + \ldots + 1.$$

Adding the terms that lie in the same vertical line we obtain

$$2S = 101 + 101 + \ldots + 101 = 100 \times 101$$

and so

$$S = \frac{100 \times 101}{2} = 5050.$$

Therefore the number that Gauss wrote on his slate should have been 5050. The method we have just described for summing an arithmetic progression is both fast and simple, and because it is simple, it is not prone to computational errors. We shall now repeat the method to obtain the more general sum

$$S_n = 1 + 2 + \ldots + n.$$

Reversing the order of the terms we obtain

$$S_n = n + (n - 1) + \ldots + 1.$$

Therefore

$$2S_n = (n+1) + (n+1) + \ldots + (n+1) = n(n+1),$$

and so we have

$$S_n = \frac{n(n + 1)}{2}.$$

This method of summing an arithmetic progression can also be displayed visually. The sum is the number of ones in the triangular array that is displayed below.

```
1
1 1
1 1 1
. . . .
1 1 1 . . . 1
```

By adding a second such triangular array and a diagonal of zeros, we obtain the square array shown below.

```
0 1 1 . . . 1
1 0 1 . . . 1
1 1 0 . . . 1
. . . . . . . .
1 1 1 . . . 0
```

The total number of entries in this array is clearly $(n+1)(n+1)$, and of these, $n+1$ of the entries are zero. The number of ones in the figure is therefore

$$(n+1)(n+1) - (n+1) = n(n+1).$$

We therefore conclude that the number of ones in the first array must be $n(n+1)/2$.

Using the result that we have just obtained we can easily find the sum of the general arithmetic progression. We start by noting that

$$d + 2d + 3d + \ldots + (n-1)d$$
$$= d(1+2+\ldots+(n-1)) = d\ \frac{(n-1)n}{2}\ ,$$

and finally we obtain

$$a+(a+d)+(a+2d)+\ldots+(a+(n-1)d)$$
$$= na+d+2d+4d+\ldots+(n-1)d$$
$$= na+d\ \frac{(n-1)n}{2}\ .$$

We see, therefore, that the summation of the special arithmetic progression

$$1 + 2 + \ldots + n$$

is the key to the general problem of summing an arithmetic progression.

1.2   Geometric Progressions

The arithmetic progressions that we studied in the preceding subsection were characterized by the property that if $a_1, a_2, \ldots, a_n$ is an arithmetic progression, then the *difference* $a_k - a_{k-1}$ between any two consecutive members is always equal to d. This is, of course, why we call d the common *difference* of the progression. If we replace these differences $a_k - a_{k-1}$ by *ratios* a_k/a_{k-1}, then the sequence $a_1, a_2, \ldots, a_n$ is said to be a *geometric progression*. In other words, a finite sequence

$$b_1, b_2, \ldots, b_n$$

of numbers is said to be a *geometric progression* if there is a number q such that for each k we have

$$\frac{b_{k+1}}{b_k} = q.$$

If the first term b_1 of this sequence is a, then we have

$$b_1 = a,$$
$$b_2 = aq \quad \text{(because } b_2/b_1 = q \text{ is tantamount to } b_2 = b_1 q = aq \text{)},$$
$$b_3 = aq^2,$$
$$\ldots$$
$$b_n = aq^{n-1}.$$

We can therefore say that a geometric progression is a finite sequence that has the form

$$a, aq, aq^2, \ldots, aq^{n-1}.$$

As we have said, the number a is the first term of the sequence. The number q is called the *common ratio* of the sequence. The problem that we shall be considering in this subsection is to find the sum of all the terms of a geometric progression, and for this purpose we shall begin by looking at the important special case that occurs when $a = 1$. In this case we have the sequence

$$1, q, q^2, \ldots, q^{n-1}.$$

We define

$$T_n = 1 + q + q^2 + \ldots + q^{n-1}.$$

Multiplying both sides of this equation by q, we obtain

$$qT_n = \quad q + q^2 + q^3 + \ldots + q^n.$$

Now subtracting and canceling terms, we obtain

$$T_n - qT_n = 1 - q^n,$$

and

$$T_n(1 - q) = 1 - q^n,$$

and as long as $q \neq 1$, we may divide by $q - 1$ to obtain

$$T_n = \frac{1 - q^n}{1 - q}.$$

In the event that $q = 1$, we simply have

$$T_n = 1 + 1 + 1 + \ldots + 1 = n.$$

We shall assume that $q \neq 1$ from now on.
Now that we have summed the special geometric progression

$$1, q, q^2, \ldots, q^{n-1},$$

we can sum the general case quite easily. Suppose that

$$T_n = a + aq + aq^2 + \ldots + aq^{n-1}.$$

Then we have

$$T_n = a(1 + q + q^2 + \ldots + q^{n-1}) = \frac{a(1 - q^n)}{1 - q}.$$

The general form of a geometric progression has some interesting applications when $q = 2$. In this case we have

$$1 + 2 + 2^2 + \ldots + 2^{n-1} = 1 + 2 + 4 + 8 + \ldots + 2^{n-1}$$
$$= \frac{1 - 2^n}{1 - 2} = \frac{2^n - 1}{2 - 1} = 2^n - 1.$$

One of these applications concerns the following story:

> An Indian king wanted to bestow a reward on a wise man who had done him a great service, and he asked the man what he desired. The wise man replied: "Take a chessboard, and put one grain of rice on the first of its 64 squares. Then put two grains on the second square and continue in this way, doubling up, until we arrive at the last square. This rice will feed my pupils and my staff." The king felt that this request was quite modest, but when they started to fulfill the wise man's wish, they found that there wasn't enough rice in the entire kingdom to do it.

In fact, the number of rice grains required to fulfil the wise man's request is $2^{64} - 1$, which is greater than the number

$$10^{18} = 1\ 000\ 000\ 000\ 000\ 000\ 000.$$

There are also some interesting applications of geometric progressions in the case $q = \frac{1}{2}$. In this case we have

$$1 + \frac{1}{2} + \frac{1}{4} + \ldots + \frac{1}{2^{n-1}} = \frac{1 - \dfrac{1}{2^n}}{1 - \dfrac{1}{2}} = 2(1 - 1/2^n) = 2 - 1/2^{n-1}.$$

This case is particularly interesting because it may be used to illustrate an important mathematical notion known as *convergence*. The following three examples illustrate the concept of convergence:

> (1) A computer is given the task of printing all of the natural numbers on a strip of paper.
> It takes 1 second to print the number 1.
> It takes $\frac{1}{2}$ second to print the number 2.
> It takes $\frac{1}{4}$ second to print the number 3.

. . .

It is possible to show that the computer will finish this task in exactly 2 seconds.

(2) A sugar bowl contains exactly 2 cubes of sugar.

The first polite guest takes 1 cube.

The second polite guest takes ½ cube.

The third polite guest takes ¼ cube.

No matter how many guests come in this way, there will always be some sugar left.

(3) One of the famous paradoxes of Zeno of Elea (about 490–430 B.C.) may be stated roughly as follows: Achilles runs a race with a tortoise, but because he can run twice as fast as she can, he allows her to start at the midpoint of the course. At the moment that he reaches her starting position, she has already arrived at the position $3/4 = 1 - 1/4$. At the moment he reaches this position, she has reached position $7/8 = 1 - 1/8$, and so on. It seems as if he will never catch up with her. On the other hand, we know that both of them will arrive at the finish line at the same moment.

All three of these examples can be explained using the notion of convergence. According to this notion, the sequence

$$1, \frac{1}{2}, \frac{1}{4}, \ldots, \frac{1}{2^n}, \ldots$$

converges to the *limit* 0. Symbolically, we write

$$\lim_{n \to \infty} \frac{1}{2^n} = 0.$$

On the other hand, the sequence whose nth member is

$$1 + \frac{1}{2} + \frac{1}{4} + \ldots + \frac{1}{2^n} = 2 - \frac{1}{2^n}$$

converges to the limit 2.

Roughly speaking, the idea of convergence of a sequence to a limit says that a sequence

$$x_1, x_2, \ldots, x_n, \ldots$$

converges to a limit x if the number x_n will be as close as we desire to the number x for all large enough values of n. An important fact about convergence is that whenever $0 \leq q < 1$, then we have

$$\lim_{n \to \infty} q^n = 0.$$

We can interpret this statement in the case $q = 0.99$ by saying that if a man starts with $100 in his bank account and withdraws one percent of his money every month, then after n months he will have $100 \cdot (0.99)^n$ dollars in the bank. As time goes on, his assets will dwindle to zero. Now since

$$1 + q + q^2 + \ldots + q^{n-1} = \frac{1 - q^n}{1 - q},$$

it can be seen that if $0 \le q < 1$, then

$$\lim_{n \to \infty} (1 + q + q^2 + \ldots + q^{n-1}) = \lim_{n \to \infty} \left(\frac{1 - q^n}{1 - q}\right) = \frac{1}{1 - q},$$

and this statement is often written more simply as

$$1 + q + q^2 + \ldots = \frac{1}{1 - q}.$$

In the three preceding examples we had $q = 1/2$, and these examples hinged upon the fact that

$$1 + \frac{1}{2} + \frac{1}{2^2} + \ldots = \frac{1}{1 - \dfrac{1}{2}} = 2.$$

1.3   The Divergence of the Harmonic Series

In the preceding subsections we considered the summation

$$1 + \frac{1}{2} + \frac{1}{2^2} + \ldots = 2.$$

In this subsection we shall consider the summation

$$1 + \frac{1}{2} + \frac{1}{3} + \frac{1}{4} + \frac{1}{5} + \ldots + \frac{1}{n} + \ldots,$$

which is known as the *harmonic series*, and we shall show that this sum is infinity. By this we mean that if we define the nth *partial sum* of our series as

$$H_n = 1 + \frac{1}{2} + \frac{1}{3} + \frac{1}{4} + \frac{1}{5} + \ldots + \frac{1}{n}$$

for each natural number n, then given any number A we can make the number H_n be greater than A by choosing n large enough:

(1) $$1 + \frac{1}{2} + \frac{1}{3} + \frac{1}{4} + \frac{1}{5} + \ldots + \frac{1}{n} > A$$

for all n from some n_0 onward (n_0 will depend upon A). This statement can be summarized by saying that the harmonic series *diverges* or that the sequence of numbers H_n *increases without bound*. To prove that the harmonic series diverges, we group the terms in the expression

$$1 + \frac{1}{2} + \frac{1}{3} + \frac{1}{4} + \frac{1}{5} + \ldots + \frac{1}{n}$$

in the following way:

$$1 + \frac{1}{2} + \left(\frac{1}{3} + \frac{1}{4}\right) + \left(\frac{1}{5} + \frac{1}{6} + \frac{1}{7} + \frac{1}{8}\right) + \left(\frac{1}{9} + \frac{1}{10} + \ldots + \frac{1}{16}\right)$$

$$+ \ldots + \left(\frac{1}{2^m + 1} + \ldots + \frac{1}{2^{m+1}}\right) + \ldots + \frac{1}{n}.$$

The first group of terms in parentheses contains $2 = 2^1$ terms and is

$$\frac{1}{3} + \frac{1}{4} > \frac{1}{4} + \frac{1}{4} = \frac{1}{2}.$$

The second group of terms in parenthesis contains $4 = 2^2$ terms and is

$$\frac{1}{5} + \frac{1}{6} + \frac{1}{7} + \frac{1}{8} > \frac{1}{8} + \frac{1}{8} + \frac{1}{8} + \frac{1}{8} = \frac{1}{2}.$$

In general, the mth group of terms in parentheses contains 2^m terms and is

$$\frac{1}{2^m + 1} + \ldots + \frac{1}{2^{m+1}} > 2^m\left(\frac{1}{2^{m+1}}\right) = \frac{1}{2}.$$

We can obviously produce as many of these groups as we like by making n large enough, and as soon as we have more than $2A$ of these groups, the inequality (1) will hold.

We shall now visualize the contrast between the two statements

$$1 + \frac{1}{2} + \frac{1}{2^2} + \ldots = 2.$$

and

$$1 + \frac{1}{2} + \frac{1}{3} + \ldots = \infty.$$

If a cavern has a width of one meter and a height of $\frac{1}{2}^n$ meters at depth n meters, then we can store rectangular boxes in this cavern with a total volume of only 2 cubic meters. See Figure III.1.1. If, on the other hand, a cavern has a width of one meter and a height

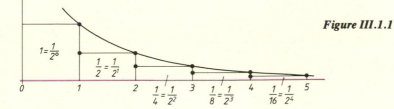

Figure III.1.1

of $1/n$ meters at depth n meters, then we can insert rectangular boxes into the cavern and then group them into infinitely many groups with a volume greater than $1/2$. See Figure III.1.2.

Another way to look at the two series is to look upon the sum

$$1 + \frac{1}{2} + \frac{1}{4} + \ldots + \frac{1}{2^n} + \ldots$$

as having been obtained from the sum

$$1 + \frac{1}{2} + \frac{1}{3} + \frac{1}{4} + \frac{1}{5} + \ldots + \frac{1}{n} + \ldots$$

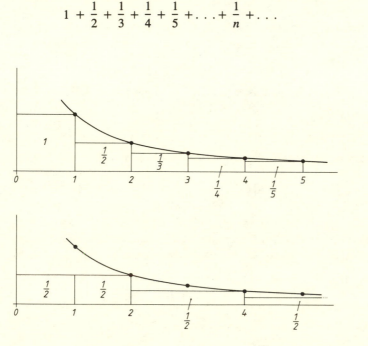

Figure III.1.2

by removing the terms that are not of the form $1/2^k$; in other words, by thinning out the harmonic series considerably. What we have seen is that this thinning process is so substantial that the sums that remain no longer increase without bound. In fact, the sums that remain are always less than 2. More generally, thinning out the harmonic series means passing from the full sequence $1, 2, 3, \ldots$ of denominators to another sequence $n_1 < n_2 < \ldots$ of natural numbers and forming

$$\frac{1}{n_1} + \frac{1}{n_2} + \frac{1}{n_3} + \ldots$$

A natural question to ask is for which thinnings of this type the partial sums of

$$\frac{1}{n_1} + \frac{1}{n_2} + \frac{1}{n_3} + \ldots$$

still increase without bound. The most famous example of this type is the sequence of all *prime numbers*

$$p_1 = 2, p_2 = 3, p_3 = 5, p_4 = 7, p_5 = 11, \ldots$$

It can be proved that

$$\frac{1}{p_1} + \frac{1}{p_2} + \frac{1}{p_3} + \ldots = \infty,$$

but a proof of this fact is too hard for us here. See Hardy-Wright [1954] and Trost [1953]. This statement about the set of prime numbers can be interpreted as saying that the primes are not too thinly scattered among the natural numbers. In contrast to this, we may say that the natural numbers of the form 2^n are very thinly scattered among the natural numbers.

1.4 Proof of the Preceding Summation Results by Mathematical Induction

In this subsection we shall show that each summation result that appears in Subsections 1.1–1.3 can also be proved by induction. In each case, we will accomplish the proof by formulating an appropriate statement A_n for each natural number n, by showing that the statement A_1 is true, and by showing that the truth of A_n implies the truth of A_{n+1} for each n.

(a) *Proof by induction that*

$$1 + 2 + 3 + \ldots + n = \frac{n(n + 1)}{2}.$$

For each natural number n we define A_n to be the statement

$$1 + 2 + 3 + \ldots + n = \frac{n(n + 1)}{2}.$$

The statement A_1 is true because it says that $1 = \dfrac{1(1 + 1)}{2}$. Now suppose that n is any natural number for which the statement A_n happens to be true. We need to show that the statement A_{n+1} holds. In other words, we need to show that

$$1 + 2 + 3 + \ldots + n + (n+1) = \frac{(n+1)\,((n+1)+1)}{2}.$$

Now we see

$$1 + 2 + 3 + \ldots + n + (n+1) = \frac{n(n+1)}{2} + (n+1) \qquad \text{(by } A_n\text{)}$$

$$= \frac{n(n+1) + 2(n+1)}{2} + \frac{(n+1)(n+2)}{2} = \frac{(n+1)\,((n+1)+1)}{2}.$$

(b) *Proof by induction that*

$$1 + q + \ldots + q^{n-1} = \frac{1 - q^n}{1 - q},$$

where $q \neq 1$. For each natural number n we define A_n to be the statement

$$1 + q + \ldots + q^{n-1} = \frac{1 - q^n}{1 - q}.$$

The statement A_1 is true because it says that

$$1 = \frac{1 - q}{1 - q}.$$

Now suppose that n is any natural number for which the statement A_n happens to be true. We need to show that the statement A_{n+1} holds. In other words, we need to show that

$$1 + q + \ldots + q^{n-1} + q^n = \frac{1 - q^{n+1}}{1 - q}.$$

Now we have

$$1 + q + \ldots + q^{n-1} + q^n = \frac{1 - q^n}{1 - q} + q^n \qquad \text{(by } A_n\text{)}$$

$$= \frac{1 - q^n + (1-q)q^n}{1 - q} = \frac{1 - q^n + q^n - q^{n+1}}{1 - p} = \frac{1 - q^{n+1}}{1 - q}.$$

(c) *Proof by induction that*

$$1 + \frac{1}{2} + \frac{1}{3} + \ldots + \frac{1}{2^n} > \frac{n}{2}.$$

For each natural number n we define A_n to be the statement

$$1 + \frac{1}{2} + \frac{1}{3} + \ldots + \frac{1}{2^n} > \frac{n}{2}.$$

The statement A_1 is true because it says that

$$1 + \frac{1}{2} > \frac{1}{2}.$$

Now suppose that n is any natural number for which the statement A_n happens to be true. We need to show that the statement A_{n+1} holds. In other words, we need to show that

$$1 + \frac{1}{2} + \frac{1}{3} + \ldots + \frac{1}{2^{n+1}} > \frac{n+1}{2}.$$

Now we have

$$1 + \frac{1}{2} + \frac{1}{3} + \ldots + \frac{1}{2^{n+1}}$$

$$= 1 + \frac{1}{2} + \frac{1}{3} + \ldots + \frac{1}{2^n} + \frac{1}{2^n + 1} + \ldots + \frac{1}{2^{n+1}}$$

$$> \frac{n}{2} + \underbrace{\frac{1}{2^n + 1} + \ldots + \frac{1}{2^{n+1}}}_{2^n \text{ terms}}$$

$$> \frac{n}{2} + \frac{2^n}{2^{n+1}} = \frac{n}{2} + \frac{1}{2} = \frac{n+1}{2}.$$

(d) *Discussion.*

Now that we have two proofs for each of the three summation theorems of this section, we need to ask ourselves some questions:

> (A) Why do we need a second proof of each statement? Was the first proof insufficient?
>
> (B) Is it really possible to have two essentially different proofs of the same statement?
>
> (C) Are the two proofs given here really different?

We shall begin with a partial answer:

> In none of the three results that we have proved is the second proof essentially different from the first one. Both proofs formulate the same basic argument, but they do so in different ways. In each case, the first proof is intuitive and the second proof is the rigorous version. The first proof is sufficient only because every mathematician knows how to transform it into the rigorous version.

You may be a little surprised to find arguments that lack rigor, in this, the most rigorous of all disciplines. As we have said, such arguments are allowed because of an expectation that any reader of the proof should be able to transform it into a completely rigorous argument. Sometimes we actually prefer the nonrigorous version. The nonrigorous version is often more intuitive and more revealing than the rigorous version, and as a result it is often easier to remember. The nonrigorous version may even provide wings to a mathematician's fantasy that enable him to perceive variations and extensions of the result that he has proved. — Let me add in passing that true rigorists don't even consider the above proofs by induction as sufficiently rigorous.

As for the question of whether there can be essentially different proofs of the same result, one is reminded of a dictum of Wittgenstein, *Every theorem is the condensed form of it own proof* (Wittgenstein [1921]). However, this statement is not consistent with the professional experience of most mathematicians. To a mathematician, a theorem is not usually an isolated entity. Rather, it is a mere building block in a much larger mathematical theory. The ideas that underlie a mathematical theory are like the architecture of a great building, but the theorems are like single bricks. If we arrive at a given mathematical theorem using two arguments that are based on theories with essentially different underlying ideas, then we would consider these two arguments as being two essentially different proofs of the theorem. We haven't given an example of this phenomenon in this section, but we shall encounter examples of this type elsewhere in this book.

In the next section we shall make more careful study of the principle of mathematical induction. At the same time, we shall learn how to recognize the lack of rigor that exists in the intuitive proofs of the three summation results that we gave in this section.

§2   *Discussion of the Principle of Induction*

In Section 1, we gave two proofs for each of the three summation theorems that we considered there. The first proof was intuitive, while the second explicitly used the principle of mathematical induction as it was introduced at the beginning of this chapter. We asserted that the induction proof was really nothing more than a more precise version of the intuitive proof, but we did not explain what this assertion means.

In this section we shall explain what we meant by that assertion. We shall begin by looking at some alternative ways of stating the principle of induction. Then we shall discuss some attempts to prove the principle. Finally, we shall explain how one may determine that a given proof is just the intuitive version of a proof by induction; and we shall explain how the induction proof should be written.

2.1   *Fundamental Observations about the Induction Principle*

We recall first the statement of the principle of induction that was given at the beginning of this chapter.

> *Induction Principle: First Form.* Suppose that for every natural number n we are given a statement A_n. Suppose that the statement A_1 is known to be true and that for every natural number n, if the statement A_n happens to be true, then the statement A_{n+1} must also be true. Then the statement A_n must be true for every natural number n.

> *Induction Principle: Second Form.* Suppose that for every natural number n we are given a statement B_n. Suppose that the statement B_1 is known to be true and that for every natural number n, if all of the statements $B_1, \ldots, B_n$ happen to be true, then the statement B_{n+1} must also be true. Then the statement B_n must be true for every natural number n.

Clearly the first form is a special case of the second. But the first form also implies the second: make the condition A_n say that all of the statements $B_1, \ldots, B_n$ are true. The third form of the induction principle makes use of the idea of a set:

> *Induction Principle: Third Form.* Suppose that M is a set of natural numbers, that the number 1 belongs to M, and that whenever a natural number n happens to belong to M, then the natural number $n + 1$ must also belong to M. Then M is the set of *all* natural numbers.

To see that the third form of the induction principle follows from the first form we need only define A_n to be the statement that the natural number n belongs to the set M. For the converse implication define M as the set of all those n for which A_n is true.

Is it impossible to *prove* the induction principle? Some have tried it, but to every attempt that has been made there are some objections. We now present some of these attempts along with some objections which may be raised against them.

Attempt I. We let n be any natural number, and we prove the statement A_n by starting at the natural number 1 and then working through the natural numbers 2, 3, . . . until we come to n. We argue as follows:

> A_1 is true.
> Since A_1 is true and A_1 implies A_2, the statement A_2 is true.
> Since A_2 is true and A_2 implies A_3, the statement A_3 is true.
> . . .
> Since A_{n-1} is true and A_{n-1} implies A_n, the statement A_n is true.

Objection. The procedure is correct only on an intuitive level. However, if we would try to take it as a formal proof, the proof becomes longer and longer as n becomes larger and larger, and we never arrive at a proof of all the statements A_n together. Some people would counter this objection with the remark that the truth of "all of the statements A_n together" is nothing but an abridged form of what was said in Attempt I, and that no method of proof could ever do more than that.

Attempt II. Assume that for some natural number n, the statement A_n is false. Then since the statement A_{n-1} implies A_n, the statement A_{n-1} must also be false. We can repeat this indirect argument again and again, until finally we have shown that the statement A_1 is false, contradicting the given condition that the statement A_1 is true. Since the assumption that some statement A_n is false leads to a contradiction, all of the statements must be true.

Objection. Same as for Attempt I.

Attempt III. Assume that for some natural number n, the statement A_n is false. Then there is a smallest one among the natural numbers n for which A_n is false; call this smallest natural number m. Since the statement A_1 is true, we cannot have $m = 1$, and therefore $m > 1$. Now since $m - 1 < m$, and since m is the smallest natural number n at which A_n is false, we see that the statement A_{m-1} is true. Therefore since the statement A_{m-1} implies A_m, we deduce that A_m is true, contradicting our choice of the number m.

Objection. This attempt makes use of some special properties of the system of natural numbers. In fact, we have assumed

> (a) that the system of natural numbers is ordered, and ordered in such a way that
> (b) the set of natural numbers has no largest member.
> (c) Every nonempty set of natural numbers has a least member.
> (d) Given any natural number $n \neq 1$, there is a largest natural number that is less than n.

These properties cannot be proved without making use of the principle of induction, and therefore Attempt III relies on a circular argument.

As a matter of fact, the induction principle cannot be proved. It becomes a necessary tool in any attempt to prove it because it is really the fundament of nearly every nontrivial statement about the natural number system, like those that were listed in the objection to Attempt III. These properties were summarized in five classical axioms known as the *Peano postulates* by Giuseppe Peano (1858–1932). Important publications on this subject are Dedekind [1888], Peano [1889], and Landau [1930].

We should mention, however, that there is an important objection to the development of the natural number system using the Peano postulates. We need to ask whether the Peano postulates are *consistent*—in other words, whether it is possible to work with these axioms without ever arriving at a contradiction. This question was first posed by David Hilbert (1862–1943) in his famous Paris lecture given on 18 August 1900 in which he posed several outstanding unsolved problems. The question of consistency of the Peano postulates was problem number 2 in that lecture. An important theorem of Kurt Gödel (1906–1978) tells us that it is impossible to prove that the Peano postulates are consistent by working inside the theory of natural numbers. Gödel showed that further tools need to be employed (Gödel [1931]). A consistency proof that uses other tools was found by Gerhard Karl Erich Gentzen (1909–1945) (see Gentzen [1936]). Most mathematicians consider Gentzen's result to be a satisfactory answer to the question of consistency of the Peano postulates.

2.2   Intuitive Proofs and Proofs by Induction

As we have said, there are many mathematical proofs that do not, on the face of it, seem to be induction proofs, but that any well-trained mathematician would instantly recognize to be induction proofs in disguise. These are the proofs which are not quite rigorous as they stand, but which could be replaced by rigorous induction proofs by anyone who wished to do so. The question we want to raise in this subsection is: How do mathematicians recognize proofs of this type, and how do they go about writing the rigorous induction proofs that are to replace them?

We are often faced with an induction proof "in disguise" when the proof contains phrases such as "for every natural number n." Often when such a phrase appears, it signifies a statement that we can call A_n and then prove by induction. If you take a second look at subsection 1.4, you will see that this is precisely what we did there.

One sure sign of an induction proof in disguise is the use of "dots" in the notation, such as $1, 2, \ldots, n$ or

$$1 + \frac{1}{2} + \frac{1}{3} + \ldots + \frac{1}{n}.$$

In truly rigorous mathematics, there is no place for a "dots" notation of this type. At this stage we have to admit that this notation does still occur now and then in Subsection 1.4, and we must therefore admit that the proofs given there are not completely rigorous. We shall amend this shortcoming in Section 7, where we shall discuss recursive definitions.

§3 *Elementary Theory of Combinatorics*

The theory of combinatorics is a particularly fertile playground for induction proofs. In this section we present some of the basic results about combinatorics.

3.1 *Words*

A *word* of length n is formed by placing n symbols down in a row, where by *symbols* we mean elements of some preassigned set that we call our *alphabet*. Alternative names for a word of length n are a *string* of length n, and an *n-tuple*. Note that the symbols that make up a word may be repeated. For example, if our alphabet is the set of letters in the ordinary alphabet of letters, then the word

wheelbarrow

is allowed. Alternatively, if we take our alphabet to be the set $\{0,1\}$ that contains the numbers 0 and 1, then we could have words such as

00, 111, 010, 1101, 010111.

Words of this type are known as *0-1 words*. More generally, if we have a given alphabet, then the words that can be formed using symbols from this alphabet are said to be *words over the given alphabet*. Apart from the alphabets we have already mentioned, the most frequently used alphabets are sets of the form

$\{1, 2, \ldots , a\}$ or $\{0, 1, 2, \ldots , a\}$,

where a is any natural number.

The elementary theory of enumerative combinatorics can be formulated in terms of statements about numbers of words, and we shall present the theory this way. However, we shall sometimes add other versions of a given result.

3.2 *The Number of Words of a Given Length*

How many 0-1-words are there with a given length n? Figure III.3.1 shows how to obtain all 0-1-words by starting with the empty word ☐ and deciding for every word already obtained whether to extend it to the right by a symbol 0 or a symbol 1. An arrow ↙ stands for an extension of the word by 0 and an arrow ↘ stands for an extension by 1. Each extension of this type doubles the number of positive words. So after n steps, we have doubled n times, and thus we have deduced the following result:

Theorem 3.1. There are precisely 2^n possible 0-1-words of length n.

The argument we have just given is, of course, nothing but the intuitive version of a proof by induction which we could write by taking A_n (for each $n = 0, 1, \ldots$) to be the statement

"There are precisely 2^n 0-1-words of length n."

If we were working with an alphabet with a symbols, we would have to multiply the

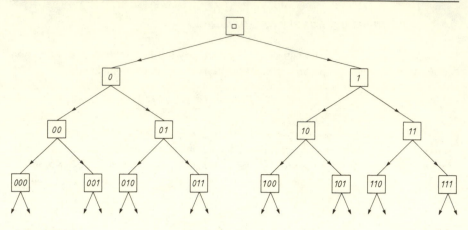

Figure III.3.1

number of possible words by *a* at every step, and we would obtain the following analogue of Theorem 3.1.

Theorem 3.2. There are precisely a^n possible words of length n over a given alphabet that has a symbols.

We can also look at the process of making a 0-1-word of length n as the process of choosing which of the n positions in the word will be occupied by the symbol 1. In all the remaining places, there will be a 0. Therefore we can restate Theorem 3.1 in the following form.

Theorem 3.3. A set of n elements has precisely 2^n subsets. Among these subsets, there is the empty set $\emptyset$ that corresponds to the word $0 \ldots 0$ that consists of nothing but zeros.

Let us suppose for the moment that our alphabet consists of the letters a, A, b, B, $\ldots$, z, Z, the space, and all the usual punctuation marks that are used in the writing of an English sentence. For words of this type we can draw a diagram similar to, but more complicated than, Figure III.3.1. Having drawn this diagram, one can draw a path through it that passes through every word exactly once. See Figure III.3.2. In this way, we have provided a list of all possible words over our chosen alphabet. This list is very extensive, for not only does it contain meaningless words like

eXib.eeta motipopA, ; skoOtch

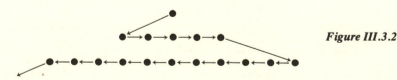

Figure III.3.2

but it also contains such words as

        To be or not to be . . . us all,

and Wycliff's Bible translation. The list even contains all books that will be written in the future. From a purely mechanical point of view, a book is nothing more than a finite sequence of symbols. Anyone with a typewriter can produce the book, simply by pressing the right buttons in the right order. Even a monkey might do it accidentally. When a set of objects can be listed in the way we have just described, we say that the set is *countable* and we therefore have the following theorem that is sometimes known as the "monkey theorem."

Theorem 3.4. The set of all finite words over a given alphabet is countable.

3.3   *The Number of 0-1-Words with a Given Number of Ones*

    In this subsection we ask how many 0-1-words of length n there can be if exactly k of the n symbols in each word are ones and the other $n - k$ symbols are zeros. Figure III.3.1 shows us how we may arrive at any given word along precisely one zigzag path of n steps ✓ and ↘. Each word has as many symbols 1 as its path contains steps ↘. See Figure III.3.3. The paths that end in level k are precisely those that end at words with k ones. The number of paths that end in row n and level k is denoted by

$$\binom{n}{k}$$

and is called the *binomial coefficient n over k*.

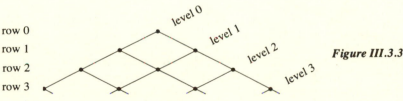

row 0
row 1
row 2
row 3

level 0
level 1
level 2
level 3

Figure III.3.3

    An important identity satisfied by binomial coefficients is

$$(1) \qquad \binom{n}{k} = \binom{n-1}{k-1} + \binom{n-1}{k}$$

which holds because $\binom{n-1}{k-1}$ is the number of paths that end in row n and level k with

↘ and $\binom{n-1}{k}$ is the number of these paths that end with ✓. We mention also that

$$\binom{n}{0} = 1 = \binom{n}{n}$$

because there can be only one path that ends in row n and level 0—namely, the path that consists only of steps ↙. In the same way we see that there can be only one path ending in row n and level n. Combining the latter identity with identity (1) we can calculate as many of the numbers $\binom{n}{k}$ in the diagram below as we like.

$$
\begin{array}{ccccccccccc}
 & & & & & 1 & & & & & \\
 & & & & 1 & & 1 & & & & \\
 & & & 1 & & 2 & & 1 & & & \\
 & & 1 & & 3 & & 3 & & 1 & & \\
 & 1 & & 4 & & 6 & & 4 & & 1 & \\
1 & & 5 & & 10 & & 10 & & 5 & & 1
\end{array}
$$

This array is known as *Pascal's triangle*. Pascal [1665] called it ''le triangle arithmétique'' and wrote it in the form shown in Figure III.3.4.

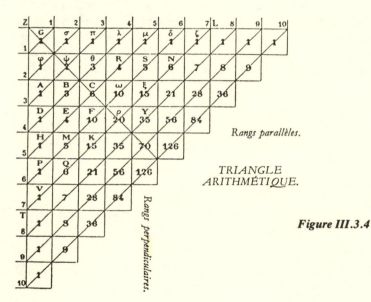

Rangs parallèles.

Rangs perpendiculaires.

TRIANGLE
ARITHMÉTIQUE.

Figure III.3.4

The triangle had been encountered previously by Stifel (1486–1567), Tartaglia (1506–1557), Stevin (1549–1620), and others, and also much earlier by the Chinese. However, Pascal was the first to make a comprehensive investigation of the number-theoretic properties of this triangle.

We now introduce factorial notation. If n is a natural number, then the symbol $n!$ (which we read as n *factorial*) is defined to be the product of all the natural numbers from 1 to n.

$$n! = 1 \cdot 2 \cdot \ldots \cdot n.$$

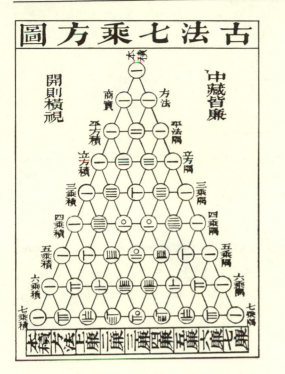

Figure III.3.5

We also define $0! = 1$. Note that $(n+1)! = (n+1) \cdot n!$ for each n. Using factorial notation, we can obtain a formula for the binomial coefficient $\binom{n}{k}$.

Theorem 3.5. If n and k are nonnegative integers and $k \le n$, then we have

$$\binom{n}{k} = \frac{n!}{k! \, (n-k)!}$$

Proof. We prove the theorem by mathematical induction. For each nonnegative integer n we define A_n to be the statement.

$$\binom{n}{k} = \frac{n!}{k! \, (n-k)!} \text{ whenever } k = 0, 1, \ldots, n.$$

The statement A_0 is true because it says that

$$\binom{0}{0} = \frac{0!}{0! 0!} \, .$$

We now show that for each natural number n, the statement A_{n-1} implies the statement A_n. Suppose that the statement A_{n-1} holds. Then, using identity (1) and this assumption, we obtain

$$\binom{n}{k} = \binom{n-1}{k-1} + \binom{n-1}{k}$$

$$= \frac{(n-1)!}{(k-1)!((n-1)-(k-1))!} + \frac{(n-1)!}{k!((n-1)-k)!}$$

$$= \frac{(n-1)!}{(k-1)!(n-k)!} + \frac{(n-1)!}{k!(n-k-1)!} \; .$$

Multiplying the top and bottom of the first of these fractions by k and the second fraction by $n-k$, we obtain

$$\frac{k(n-1)!}{k!(n-k)!} + \frac{(n-k)(n-1)!}{k!(n-k)!} = \frac{n!}{k!(n-k)!} \; .$$

This formula for $\binom{n}{k}$ looks quite elegant, but it has some disadvantages compared

with Pascal's triangle. For example, it is clear from the way Pascal's triangle is built that every entry in the triangle is an integer. On the other hand, it is not immediately obvious that the expression

$$\frac{n!}{k! \, (n - k)!}$$

is an integer. Using this formula we can now restate some of our earlier results:

Theorem 3.6. Among the 2^n possible 0-1-words of length n, there are precisely

$$\binom{n}{k} = \frac{n!}{k! \, (n-k)!}$$

words with exactly k symbols 1.

Theorem 3.7. For every natural number n we have

$$2^n = \binom{n}{0} + \binom{n}{1} + \ldots + \binom{n}{n}.$$

As we have said, we can also look at the process of making a 0-1-word of length n as the process of choosing the subset of all those among our n given places that will be occupied by the symbol 1. Thus:

Theorem 3.8. In every set of n elements, there are exactly

$$\binom{n}{k} = \frac{n!}{k! \, (n-k)!}$$

subsets that contain exactly k elements. Theorem 3.8 may also be stated in the following form:

Theorem 3.9. A set of n elements can be split into a k-element set and an $(n-k)$-element set in exactly

$$\binom{n}{k} = \frac{n!}{k! \, (n-k)!}$$

ways.

Once again we could ask how our results would be affected if we were to replace the set $\{0,1\}$ by an arbitrary finite alphabet. This leads to what are known as *multinomial coefficients*, but we shall not go into details on this topic.

3.4   The Number of Words with All Different Letters

In this subsection we shall restrict our attention to the words over a given alphabet that contain no repetitions among their letters, and we shall ask how many words of this type there can be of a given length. Theorem 3.10 effectively answers this question.

Theorem 3.10 Over an alphabet with a symbols, there exist precisely

$$a \cdot (a-1) \cdot \ldots \cdot (a-n+1)$$

repetition-free words of length n (where $n \le a$).

Proof. We may assume that the alphabet is the set $\{1, 2, \ldots, a\}$. We split the set of all repetition-free words of length n into a classes:

K_1 contains all words that begin with 1.

. . .

K_a contains all words that begin with a.

The class K_1 consists of all words of the form

$$1 \, x_1 \, x_2 \, \ldots \, x_n,$$

where $x_1 x_2 \ldots x_n$ is a repetition-free word of length $n - 1$ over the alphabet $\{2, \ldots, a\}$. We see that this alphabet contains $a - 1$ elements. From these observations we see how to set up a proof of our theorem by induction. When $n = 1$, we have a repetition-free words of length n; and therefore the theorem is true when $n = 1$. If we have already worked through $n - 1$ steps of the proof, then the nth step can be completed as follows: Each of the classes $K_1, \ldots, K_a$ contains exactly

$$(a-1) \cdot (a-2) \cdot \ldots \cdot ((a-1)-(n-1)+1) = (a-1) \cdot \ldots \cdot (a-n+1)$$

words, and so the total number of words in all a of these classes is

$$a \cdot (a-1) \cdot \ldots \cdot (a-n+1).$$

The theorem therefore follows by mathematical induction.

The special case of Theorem 3.10 that results when $n = a$ is of particular importance.

Theorem 3.11. There are exactly $n!$ ways of placing n objects in n preassigned places.

This result suggests another proof of Theorem 3.9 (and, therefore, of Theorems 3.6

and 3.7 as well) because we can achieve any distribution of n objects over n places as follows:

> Split the set of n objects into a subset of k objects and a subset of $n - k$ objects. There are $\binom{n}{k}$ = ways to do this.

> Arrange the k objects from the first subset on the first k of the n places. There are $k!$ ways to do this.

> Arrange the $n - k$ objects from the second subset on the remaining $n - k$ places. There are $(n - k)!$ ways to do this.

From these observations we see that

$$n! = \binom{n}{k} k!(n-k)! \ ,$$

and therefore

$$\binom{n}{k} = \frac{n!}{k!(n-k)!} \ ,$$

which proves Theorem 3.6 once more.

§4   *The Marriage Theorem*

The marriage theorem is a combinatorial result that tells us that under certain conditions it is possible to marry off all the women in a given society. Before we state the theorem, we should mention that we assume that the society we are describing is monogamous. There is also a "harem" version of the theorem, but we are not concerned with it here. The statement of the marriage theorem is as follows:

Theorem 4.1. Suppose that there is a system of friendships among d women and h men, and suppose that the following condition, which we call the *party condition*, holds:

> Whenever any group of women invite all their friends to a party, there will be at least as many men at the party as women.

Then it is possible for every one of the d women to marry a man who is one of her friends.

Proof. In order to prove the theorem by induction on the number d of women, we shall call the statement of the theorem A_d. We observe first that the statement A_1 is true, for if there is only 1 woman, then the party condition guarantees that she has at least one friend, and she can marry him. Now suppose that all of the statements $A_1, A_2, \ldots, A_{d-1}$ are true. In order to establish the truth of the statement A_d, we shall consider two cases.

Case I: We assume that the party condition holds in the somewhat stronger sense that whenever a group of fewer than d women invite all their friends to a party, then the number of men at the party will be at least one more than the number of women. In this case we select any one woman, and noting that she has at least two friends (why?), we marry her off to one of her friends (call him Harry) and send the happy couple off on their honeymoon. Now suppose that any group of the remaining women invite their friends to a party. If Harry could have been present at this party, we would know that the number of men at the party is at least one more than the number of women. Therefore, even though Harry is not present, we are assured that there are as many men at the party as there are women. We observe, therefore, that the people who remain satisfy the party condition, and so it is possible to marry off the remaining $d - 1$ women to their friends.

Case II: We assume that it is possible to find a group of (say) d_0 women, where $d_0 < d$, such that when these particular women invite their friends to a party, there will be exactly d_0 men at the party. We now apply condition A_{d_0} to this group of women to marry them off and send them and their husbands off on their honeymoon. Now among the $d_1 = d - d_0$ women who are left, if any group of, say, e of these women invite all their still unmarried friends to a party, then there will be at least e men at the party. The reason for this is that if there were fewer than e men at this party, then since the d_0 newly married women can contribute only their husbands, a party organized by all $d_0 + e$ women—the e unmarried ones plus the d_0 married ones—would have fewer than $d_0 + e$ men in it. We may therefore apply condition A_{d_1} to the d_1 women who remain, and marry them off.

The theorem therefore follows by induction.

A question related to the marriage theorem is how many ways there may be for marrying off all the women. Partial answers to this question are given by Ryser [1963] (see also Jacobs [1969b, 1983a]). We shall encounter an entirely different proof of the marriage theorem in Section 2 of Chapter IV.

§5   *The Binomial Theorem*

By looking at Pascal's triangle we see at once that the well-known identities

$$(a+b)^0 = 1$$
$$(a+b)^1 = a+b$$
$$(a+b)^2 = a^2 + 2ab + b^2$$
$$(a+b)^3 = a^3 + 3a^2b + 3ab^2 + b^3$$

can be written in the form

$$(a+b)^0 = \binom{0}{0} a^0 b^0$$

$$(a+b)^1 = \binom{1}{0} a^1 b^0 + \binom{1}{1} a^0 b^1$$

$$(a+b)^2 = \binom{2}{0} a^2 b^0 + \binom{2}{1} a^1 b^1 + \binom{2}{2} a^0 b^2$$

$$(a+b)^3 = \binom{3}{0} a^3 b^0 + \binom{3}{1} a^2 b^1 + \binom{3}{2} a^1 b^2 + \binom{3}{3} a^0 b^3.$$

The binomial theorem is the statement that this sequence of identities can be continued to include $(a+b)^n$ for every natural number n. We shall prove the theorem by induction.

Theorem 5.1 (The Binomial Theorem). Given any numbers a and b and any natural number n, we have

$$(a+b)^n = \binom{n}{0}a^n b^0 + \binom{n}{1}a^{n-1}b^1 + \ldots + \binom{n}{n-1}a^1 b^{n-1} + \binom{n}{n}a^0 b^n.$$

Proof. In order to prove the theorem by induction, we call the required identity A_n. We have already seen that the statement A_1 is true. Now suppose that n is any natural number for which the statement A_{n-1} is true. To see that the statement A_n is also true, we note that

$$(a+b)^n = (a+b)^{n-1}(a+b)$$

$$= \left(\binom{n-1}{0}a^{n-1}b^0 + \ldots + \binom{n-1}{n-1}a^0 b^{n-1} \right)(a+b)$$

$$= \binom{n-1}{0}a^{n-1}b^0 a + \ldots + \binom{n-1}{k}a^{n-1-k}b^k a + \ldots + \binom{n-1}{n-1}a^0 b^{n-1}a$$

$$+ \binom{n-1}{0}a^{n-1}b^0 b + \ldots + \binom{n-1}{k-1}a^{n-1-(k-1)}b^{k-1}b + \ldots + \binom{n-1}{n-1}a^0 b^{n-1}b$$

$$= \binom{n-1}{0}a^n b^0 + \binom{n-1}{1}a^{n-1}b^1 \ldots + \binom{n-1}{k}a^{n-k}b^k + \ldots + \binom{n-1}{n-1}a^1 b^{n-1}$$

$$+ \binom{n-1}{0}a^{n-1}b^1 + \ldots + \binom{n-1}{k-1}a^{n-1-(k-1)}b^k + \ldots + \binom{n-1}{n-1}a^0 b^n.$$

Now using the facts that

$$\binom{n-1}{0} = 1 = \binom{n}{0} \quad \text{and} \quad \binom{n-1}{n-1} = 1 = \binom{n}{n}$$

we obtain

$$\binom{n}{0}a^n b^0 + \left[\binom{n-1}{1} + \binom{n-1}{0} \right]a^{n-1}b^1$$

$$+ \ldots + \left[\binom{n-1}{k} + \binom{n-1}{k-1} \right]a^{n-k}b^k$$

$$+ \ldots + \left[\binom{n-1}{n-1} + \binom{n-1}{n-2} \right]a^1 b^{n-1} + \binom{n}{n}a^0 b^n,$$

and using the identity

$$\binom{n}{k} = \binom{n-1}{k-1} + \binom{n-1}{k}$$

upon which we based Pascal's triangle, we obtain finally

$$\binom{n}{0}a^n b^0 + \binom{n}{1}a^{n-1}b^1 + \ldots + \binom{n}{n-1}a^1 b^{n-1} + \binom{n}{n}a^0 b^n.$$

The name "binomial theorem" is derived from the term "binomial" that is used to describe an algebraic expression that has two terms, like $a + b$.

§6   *Induction Proofs of Two Fundamental Theorems*

In this section we shall prove two important results by induction: the well ordering of the system of natural numbers, and a result known alternatively as the *pigeonhole principle* or the *Dirichlet drawer principle*.

6.1   *The Well-Ordering of the System of Natural Numbers*

In describing our objections to the third "proof" of the induction principle that was given in Subsection 2.1, we pointed out that this proof made use of the assumption that every nonempty set of natural numbers has a least member. We indicated at the time that this property of the system of natural numbers could itself be deduced from the principle of induction and that if we use each of these statements to prove the other, we are arguing in a circle. In this subsection we show that if we assume the principle of induction, then indeed we can prove that every nonempty set of natural numbers has a least member.

In general, if $\leq$ is an order (alternatively known as a *total order*) in a set M, then $\leq$ is said to be a *well order* of M if every nonempty subset of M has a least member. Therefore our task in this subsection is to show that (assuming the induction principle) the system of natural numbers is well ordered.

Theorem 6.1. Every nonempty set of natural numbers contains a smallest member.

Proof. For each natural number n we define A_n to be the statement that if a set M of natural numbers contains n, then M has a least member. We observe first that the statement A_1 is true because any set M that contains the number 1 must have 1 as its least member. Now suppose that n is any natural number with the property that the statement A_m is true for every natural number $m \leq n$. Let M be any set of natural numbers containing the number $n + 1$. We need to show that M has a least member. We consider two cases:

Case I: Suppose that there is a natural number $m \leq n$ such that m is contained in the set M. By statement A_m, the set M has a least member.

Case II: Suppose that there is no natural number $m \leq n$ such that m is contained in M. In this case, the number $n + 1$ is clearly the least member of M.

6.2   *The Pigeonhole Principle and Dedekind's Definition of a Finite Set*

Roughly speaking, the pigeonhole principle says that if more than n pigeons have to share n pigeonholes, then at least one of the holes must be occupied by more than one pigeon. The pigeonhole principle is sometimes known as the Dirichlet drawer principle (even though it was never published as such by Dirichlet), and in this form it says that if more than n items are stored in n drawers, then at least one drawer must contain more than one item.

As a further illustration of the pigeonhole principle we might say that if a factory employs more than 366 workers, then at least two of the workers must have the same birthday.

In order to prove the pigeonhole principle by induction, we define A_n for each natural number n to be the statement

> *If more than n items are stored in n drawers, then at least one drawer must contain more than one item.*

The statement A_1 is clearly true. Now suppose that n is any natural number for which the statement A_n is true. We need to show that A_{n+1} is true. Suppose that more than $n+1$ items are stored in $n+1$ drawers. We now open one of the drawers. There are two cases.

Case I: If the drawer we have opened contains more than one item, the proof is complete.

Case II: Suppose that the drawer that we have opened contains just one item. Then since the remaining n drawers must hold more than n items, we deduce from statement A_n that one of these remaining drawers must contain more than one item.

It is worth noting that the pigeonhole principle is an *existence statement*. In the form that we have just proved it, it tells us that a drawer *exists* that contains more than one item. However, the proof does not tell us how to find this drawer. Still another way of stating the pigeonhole principle is as follows:

> Suppose that we have a number of items that need to be stored in n drawers and that every drawer contains at least one item. Then it is impossible to empty any drawer, moving its content to other drawers, without causing at least one other drawer to contain more than one item.

This statement becomes more relevant to everyday life if we replace "drawer" by "hotel room" and "item" by "guest." Now we shall look at this form of the principle in more abstract terms.

> Suppose that M is a finite set. It is impossible to assign each member of M to another member of M in one-to-one fashion in such a way that there is a member of M to which no member of M has been assigned.

Note that when we say that we are assigning members of M to other members of M in a one-to-one fashion, we mean that no two members of M are assigned to the same member. Putting this statement in the familiar language of functions, we obtain it in the following form.

Suppose that M is a finite set. It is impossible to find a one-to-one function from M into a proper subset of M.

In his famous book, *Was sind und was sollen die Zahlen* (What numbers are and what they are good for, Dedekind [1888]), Richard Dedekind (1831–1916) *defined a finite* set to be a set that has this property. Dedekind based his approach to the system of natural numbers upon this definition, which is known as the *Dedekind definition of finiteness*. Dedekind's approach is essentially equivalent to that of Peano [1887].

6.3   Ramsey's Theorem

The pigeonhole principle can be generalized in several important ways. For example, the following sharpened form of the principle is more specific about the number of items that must be in the same drawer.

If at least $nk + 1$ items are stored in n drawers, then at least one drawer must contain at least $k + 1$ items.

Another, much more profound generalization of the principle was given in 1930 by Frank Plumpton Ramsey (1903–1930) (see Ramsey [1930], Ryser [1963], and Jacobs [1983a]). Ramsey's theorem is too complicated to be presented here in full detail, but we shall display one of its applications.

Suppose that N points are placed in the plane. Since any three of these points determine a triangle, there are $\binom{N}{3}$ such triangles. Suppose that we now split these triangles into two classes, color the triangles in the first class white, and color the triangles in the second class black. We shall now call a set containing n of the original N points *monochrome* if all $\binom{n}{3}$ triangles formed by triples from these n points have the same color.

Ramsey's theorem allows us to conclude that there is a number $N(2,n)$ such that for every $N \geq N(2,n)$ and every possible coloring of the $\binom{N}{3}$ triangles, there will be a monochrome set with at least n points.

Roughly speaking, the theorem tells us that if we put enough points in the plane, then we can be sure that there are arbitrarily large monochrome subsets, no matter how we color the triangles obtainable out of them.

As you may have guessed, Ramsey's theorem is not restricted to triangles. We could have stated this result in terms of k-point subsets, where k is any natural number, and in fact the theorem is proved by induction on the number k. In the event that $k = 1$, Ramsey's theorem reduces to the pigeonhole principle. Like the pigeonhole principle, Ramsey's theorem is an existence statement. It tells us that large monochrome subsets exist, and not how to find them. The theorem has led to a well-established branch of mathematics known as "Ramsey theory" that contains the theorems of van der Waerden and Szemerédi that we shall describe in Subsection 7.2. Incidentally, the numbers $N(2,n)$, which are called Ramsey numbers, are so large that they cannot be handled by the methods of ordinary arithmetic (Paris-Harrington [1977]).

§7   *Recursive Definition*

The principle of induction is a *method of proof*, but, along with it, there is a method of "inductive definition" that we call *recursion* or *recursive definition*. Intuitively speaking, recursive definition is the process of defining a sequence of objects g_0, g_1, g_2, . . . in a step-by-step fashion using g_{n-1} at each stage as a basis for defining g_n. The logical foundation for definitions of this type is provided by the following theorem.

Theorem 7.1. Suppose that G_0, G_1, . . . are given sets, and that for each $n = 0, 1, . . .$ we are given a function K_n that associates to any member x of G_n, some unique member $K_n(x)$ of the set G_{n+1}. Then given any element g_0 of the set G_0, there is a unique sequence g_0, g_1, . . . such that for every $n = 0, 1, . . .$ we have

(a) g_n is a member of the set G_n.
(b) $g_n = K_n(g_{n-1})$.

Intuitively speaking, we would try to prove this theorem by starting with g_0, then defining $g_1 = K_0(g_0)$, then defining $g_2 = K_1(g_1)$, and continuing in this fashion. A rigorous proof based on the principle of induction would have to be more technical than we would like to afford here; we omit it. Using the method of recursive definition, we can eliminate the dots . . . from our notation and thus finally fulfill our promise given at the end of Section 2. The following examples show how this may be done.

(1) The expression $1 + 2 + . . . + n$ can be written in the form

$$\sum_{k=1}^{n} k.$$

This symbol may be recursively defined by requiring that

$$\sum_{k=1}^{1} k = 1 \qquad \text{and} \qquad \sum_{k=1}^{n+1} k = \left[\sum_{k=1}^{n} \right] + (n+1).$$

In the same way we can write other summations such as

$$1 + \frac{1}{2} + \frac{1}{3} + . . . + \frac{1}{n} = \sum_{k=1}^{n} \frac{1}{k}$$

and

$$1 + q + q^2 + . . . + q^{n-1} = \sum_{k=0}^{n-1} q^k.$$

The "summation index" k can be replaced by the symbol j, or by any other symbol we like, and it can run from any integer m we like to any integer n $\geq m$. It can readily be seen that the expression

$$\sum_{k=1}^{n} 1$$

stands for $1 + 1 + \ldots + 1$ (n times), and so

$$\sum_{k=1}^{n} 1 = n.$$

This summation notation can also be extended to infinite sums. For example,

$$1 + \frac{1}{2} + \frac{1}{3} + \ldots = \sum_{n=1}^{\infty} \frac{1}{n}.$$

(2) By analogy with the way we wrote sums in the preceding example, one may also write a product $c_1 c_2 \cdot \ldots \cdot c_n$ of n numbers $c_1, c_2, \ldots, c_n$ in the form

$$c_1 \cdot c_2 \cdot \ldots \cdot c_n = \prod_{k=1}^{n} c_k = \prod_{i=1}^{n} c_i.$$

This expression is defined recursively as follows:

$$\prod_{k=1}^{1} c_k = c_1 \text{ and } \prod_{k=1}^{n+1} c_k = \left[\prod_{k=1}^{n} c_k \right] \cdot c_{n+1}.$$

(3) Even complicated expressions such as

$$1 + \frac{1}{2} + \left(\frac{1}{3} + \frac{1}{4} \right) + \left(\frac{1}{5} + \frac{1}{6} + \frac{1}{7} + \frac{1}{8} \right) + \ldots +$$

$$\left(\frac{1}{2^n + 1} + \ldots + \frac{1}{2^{n+1}} \right)$$

can be written without the "dots" using recursively defined expressions. The latter expression can be written as

$$1 + \frac{1}{2} + \sum_{k=1}^{n} \sum_{j=1}^{2^k} \frac{1}{2^k + j}.$$

Strictly speaking, Pascal's triangle and the infinite diagrams that we discussed in Section 2 should also be defined recursively, and so should any of the paths that we drew through the diagrams in Section 2.

Even the process of writing down the sequence $1, 2, \ldots$ of all natural numbers looks like a recursive definition. However, the basis for the idea of recursive definitions itself depends upon the system of natural numbers, and so if we were to use recursion to construct the sequence of natural numbers we would be arguing in a vicious circle. On the other hand, from a practical or "pre-mathematical" perspective, we could consider the sequence of natural numbers a succession of rows of strokes, $|, \ |\,|, \ |\,|\,|, \ \ldots$, and regard this sequence as some sort of primeval recursion. In this case, the recursion would be as follows:

> Begin with no stroke ($n = 0$).
> At each stage n, add one more stroke.

Instructions of this type can be fed into a computer, which will then implement the process and print out the sequence until we run out of memory, paper, money, time, or patience.

In the final three subsections of this section, we shall look at some recursively defined sequences that are of particular importance.

7.1   The Fibonacci Numbers

We begin with a little history of this topic.

Fibonacci (1180–1250) was the most important European mathematician of the high medieval era. His true name was Leonardo of Pisa, but his father, a merchant who was active in such places as Algiers, had been given the nickname "Bonaccio": the good-hearted one. Therefore, his son was called "Fi-Bonacci," the son of Bonaccio. Bonaccio recognized his son's scientific talent when Fibonacci was quite young, and he sent him around the Islamic world. No doubt, some of these travels were to promote his father's business, but they also had the effect of introducing the young Fibonacci to the state of the art in Indo-Arabian mathematics. As a result of these travels, Fibonacci published the book *Liber Abaci* (The book of the abacus), which appeared in its first edition in 1202. The systematic use in western Europe of the Arabian system of numerals came about as a direct result of this book. Fibonacci's contributions dominated European mathematics over the next three centuries. His genius was soon recognized. Friedrich II of Hohenstaufen, the great "intellectual emperor," once arranged a public discussion between Fibonacci and his own court mathematician, a discussion that featured many intricate calculations that were carried out mentally. Friedrich II instructed Fibonacci to write a second edition of *Liber Abaci*, which he did in the year 1228. It is on this edition that the critical edition of 1857 is based and in which the following passage appears (page 283).

Quot para coniculorum in uno anno ex uno pario germinantur.	How many pairs of rabbits will be generated by one pair in one year?
Quidam posuit unum par cuniculorum in quodam loco, qui erat undique pariete circundatus, ut sciret, quot ex eo pario germinarentur in uno anno:	Someone confined a pair of rabbits to an area enclosed by walls from all sides. He wanted to find out how many pairs would be generated, starting with this pair, in one year.
cum natura eorum sit per singulum mensem aliud par germinare; et in secundo mense ab eorum nativitate germinant.	It is the nature of rabbits to give birth to one new pair every month; and they start this at the age of two months.

Quia suprascriptum par in primo mense germinat, duplicabis ipsum, erunt paria duo in uno mense.

The said pair starts proliferating right away. Thus after one month we have two pairs.

Ex quibus unum, scilicet primum, in secundo mense germinat; et sic sunt in secundo mense paria 3; ex quibus in uno mense duo pregnantur; et germinantur in tercia mense paria 2 coniculorum; et sic sunt paria 5 in ipso mense; ex quibus . . .

Of these, the original pair gives birth to another pair next time. This gives a total of three pairs after 2 months. Of these three, two give birth in the next period; five pairs after 3 months.

Fibonacci continues in this fashion until the twelfth month, when there are 377 pairs of rabbits. On the margin of that page in his book, he displays a little table that contains the numbers

$$1,2,3,5,8,13,21,34,55,89,144,233,377$$

Fibonacci ends the section with the remark

et sic posses facere per ordinem de infinitis numeris mensibus.

and in this fashion we could go on step by step to an arbitrary number of months.

If we translate Fibonacci's work into modern language, we obtain an infinite sequence $F_0, F_1, \ldots$ The numbers in this sequence are known as *Fibonacci numbers*. In constructing the sequence today we assume that the original pair of rabbits begins proliferating only after two months.

> At time 0 there is one pair of rabbits. Therefore $F_0 = 1$.
> At time 1 there is still only one pair of rabbits. Therefore $F_1 = 1$.
> At time 2 there is the original pair and 1 newborn pair. Therefore $F_2 = 2$.
> At time $n + 1$ there are the F_n pairs from the previous month plus F_{n-1} newborn pairs. Therefore

(1) $F_{n+1} = F_n + F_{n-1}$.

This identity is the *recursion formula* for the Fibonacci numbers. Using it we can calculate the Fibonacci numbers step by step as far as we like, starting from the *initial conditions*

(2) $F_2 = F_1 = 1$.

Fibonacci displayed this table:

n	0	1	2	3	4	5	6	7	8	9	10	11	12	13
F_n	1	1	2	3	5	8	13	21	34	55	89	144	233	377

The Fibonacci recursion is the oldest known contribution to a flourishing field of modern biomathematics known as *population dynamics*. Fibonacci's work in this field is so far-reaching that an entire journal, "The Fibonacci Quarterly," has been devoted to this field since 1963. one of the most striking results about the Fibonacci numbers is the following formula, which was discovered by de Moivre in 1718.

$$(3) \qquad F_n = \frac{1}{\sqrt{5}} \left[\left(\frac{1+\sqrt{5}}{2} \right)^{n+1} - \left(\frac{1-\sqrt{5}}{2} \right)^{n+1} \right].$$

It is intriguing to find the number $\sqrt{5}$ here again. The last time we saw this number was in Chapter I when we were discussing the golden section. The number $\sqrt{5}$ is not even rational, but in spite of this, the right side of the identity (3) is always a natural number because it is the number F_n. On closer inspection we can see that for any given value of n, the $\sqrt{5}$ either cancels out or is squared, in either case leaving us with a natural number. For example,

$$F_1 = \frac{1}{\sqrt{5}} \left[\left(\frac{1+\sqrt{5}}{2} \right)^2 - \left(\frac{1-\sqrt{5}}{2} \right)^2 \right]$$

$$= \frac{1}{\sqrt{5}} \left(\frac{1+2\sqrt{5}+(\sqrt{5})^2}{4} - \frac{1-2\sqrt{5}+(\sqrt{5})^2}{4} \right) = 1.$$

Of course, this value coincides with that in the table given above. (See also Jacobs [1983a].) This example illustrates how the $\sqrt{5}$ is eliminated in the general case. We notice that the recursion formula (1) would lead to integers even if the initial conditions were changed, provided that the new initial values are integers.

7.2   The Thue-Morse Sequence

It is possible to produce interesting mathematics without mentioning numbers explicitly. If Napoleon (1769–1821) could have been advised by the Norwegian mathematician Axel Thue (1863–1922), or the American mathematician Marston Morse (1892–1977), when he planted his rows of poplars all over Europe, these rows of poplars could have resembled Figure III.7.1. We can imagine that these rows of trees could have

Figure III.7.1

given rise to wild speculations on the part of a Martian astronaut who had come to explore the earth. The law according to which this row of trees is formed becomes transparent the moment we look at initial sections of length $1 = 2^0, 2 = 2^1, 4 = 2^2, 8 = 2^3, \ldots$

Napoleon plants the first tree 🌳 . Ranger number 1 plants the "opposite" kind of tree 🌳 . Ranger number 2 finds 🌳🌳 and plants the opposite 🌳🌳 . Ranger number 3 finds 🌳🌳🌳🌳 and plants the opposite 🌳🌳🌳🌳 etc. See Fig. III.7.2.

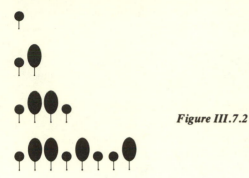

Figure III.7.2

So the row of trees in Figure III.7.1 is planted according to the principle: Der Geist der stets verneint (the spirit who always says no). On the other hand, "the spirit who always says yes first and then says no" would have planted the row of trees shown in Figure 7.7.3.

Figure III.7.3

Mathematicians replace ♀ by 0 and ♀ by 1, declaring 0 and 1 to be opposites. In this way they arrive at the following sequences:

        01101001. . .   ("The spirit who always says no")
        001001110. . .   ("Mephisto Waltz")

Both of these sequences abound in symmetries. If we partition the first sequence into blocks of length 2 we obtain 01|10|10|01|. . . , and if we now replace 01 by 0 and 10 by 1, we obtain the original sequence 01101001. . . again. Objects (for example, geometric objects) that have this kind of symmetry are called *fractals* nowadays and are currently the subject of intensive research. There have been suggestions that this research may throw light on the structure of matter, and even on the origin of life (see, for example, Mandelbrot [1977]). Such far-flung consequences were far from the minds of Axel Thue, who discovered the sequence 01101001. . . in 1904, and Marston Morse, who rediscovered it independently in 1921. Thue was looking for a sequence in which certain repetition phenomena would *not* occur. In fact Hedlund-Morse [1944] proved that blocks of the form

$$b_0 b_1 \ldots b_n b_0 b_1 \ldots b_n b_0$$

which contain a block of characters, a repetition of this block, and then the first character of the block, do *not* occur in the sequence 01101001. . . Examples of such impossible strings are 000, 111, 01010. Gottschalk [1964] has shown that the sequence

01101001. . . is, up to minor modifications, the only sequence that has the latter property (see, for example, Jacobs [1983a], Dekking [1979]). Precise proofs of these statements are beyond the scope of this book.

We shall confine ourselves to the proof of just one typical property of the sequence 01101001. . . We shall show that if some block B of symbols occurs in the sequence 01101001. . . at all, then it occurs with bounded gaps. What this means is that we can portion the sequence 01101001. . . into blocks $B_0, B_1, B_2,$. . . of equal length such that every one of the blocks B_n includes a copy of the block B somewhere. This does not mean that the block B must occur periodically. Mathematicians say that the block B must occur *almost periodically*. We mention in passing that the notion of almost periodicity is of fundamental importance in such fields as celestial mechanics.

The proof of this fact runs as follows: Locate the block B somewhere in the sequence 01101001. . ., and portion the sequence into blocks $A_0, A_1,$ . . . of length 2^n such that the located block B is included in the block A_0 (we need only choose n large enough to ensure this). See Figure III.7.4.

From the inner symmetry of the sequence 01101001. . . that we have mentioned, we deduce that if we define $A = A_0$ and $\bar{A} = A_1$, then the sequence 01101001. . . can be written as $A\bar{A}\bar{A}A\bar{A}AA\bar{A}$ . . . See Figure III.7.5. If we now define $C_0 = A\bar{A}$, $C_1 = \bar{A}A$, . . . , then every block $C_0, C_1,$ . . . contains a copy of block A, and hence a copy of block B, which is what we wanted to prove.

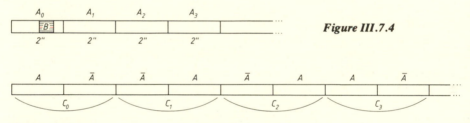

Figure III.7.4

Figure III.7.5

The reader is invited to test his understanding of this proof by showing that the "Mephisto Waltz" 001001110. . . is also almost periodic.

We take this opportunity to describe, as an appendix to this chapter, some of the deepest and most profound mathematical results of this century. We shall say that in an infinite 0-1-sequence (like 001010. . .), the symbol 1 occurs in an *arithmetical progression of length n* if we can extract, from the infinite sequence, a finite figure of the type shown in Figure III.7.6. In the same way we define the occurrence of 0 in an arithmetical progression of length n.

Theorem 7.2. (van der Waerden [1927]). In every infinite 0-1-sequence, at least one of the two symbols 0 and 1 must occur in arithmetical progressions of every length.

Nowadays this theorem is seen as a part of the Ramsey theory that we discussed in Section 6. Its proof takes the form of an intricate iterated application of the pigeonhole principle (Theorem 6.2), and is far beyond the scope of this book (see Jacobs [1983a]). The theorem can be seen to be plausible if we try to construct a 0-1-sequence in which 0

...1....1....1....1.----------.1....1....1...   *Figure III.7.6*

n times equidistant

does *not* occur in arithmetical progressions of arbitrary length. One obvious method to achieve such a sequence would be to insert longer and longer gaps between the symbols 0 that are present. Since these gaps are made up only of the symbol 1, we see that the symbol 1 can be found in arbitrarily long arithmetic progressions which have common difference 1.

Van der Waerden's theorem is purely an existence theorem. It doesn't tell us *which* of the symbols 0 and 1 will occur in arithmetic progressions of arbitrary length. This question is partially answered by the following result.

Theorem 7.3. (Szemerédi [1975]). Given a 0-1-sequence such that there exists an $\epsilon > 0$ such that for infinitely many initial sections of the given sequence, the proportion of symbols 1 exceeds ϵ, then the symbol 1 occurs in arithmetical progressions of arbitrary length.

Since the proportions of 0 and 1 in any initial section must always add up to 1, at least one of the symbols 0 and 1 must have the property of "occurrence with positive upper density" that was formulated (for the symbol 1) in the statement of Szemerédi's theorem. Therefore van der Waerden's theorem is a consequence of Szemerédi's theorem. This fact does not, however, lead to a new proof of van der Waerden's theorem, because this theorem is used in the proof of Szemerédi's theorem. Szemerédi's theorem is among the most difficult results of present-day mathematics. Details of the research that preceded it and the work that has come after it can be found in Jacobs [1983a, 1983b].

7.3   *Georg Cantor's Diagonal Method*

George Cantor (1845–1918) published his famous diagonal method in 1891 (Cantor [1891]). It may, however, be tracked already in Du Bois–Reymond [1876]. The diagonal method is applied to a sequence of 0-1-sequences. Below we show a sequence of 0-1-sequences. Each row represents one of the given 0-1-sequences. The elements that are placed in boxes lie on what we call the *diagonal*.

```
[0] 0  1  0  0  0  1  0  0  0 ...
 1 [0] 1  0  1  0  1  0  1  0 ...
 0  1 [1] 0  1  0  0  1  1  0 ...
 0  0  1 [0] 0  1  1  1  0  0 ...
 1  1  1  1 [1] 1  1  1  1  1 ...
 1  0  0  1  0 [1] 1  0  0  1 ...
 0  0  0  0  0  0 [0] 0  0  0 ...
 1  0  1  1  0  1  1 [1] 0  1 ...
 0  1  0  1  0  1  0  1 [0] 1 ...
 1  0  0  1  0  0  1  0  0 [1] ...
 .  .  .  .  .  .  .  .  .  .  .  .  .  .  .  .  .
```

Cantor's diagonal method is the process of taking the diagonal elements one by one and changing every 1 to 0 and every 0 to 1. For example, for the sequence of 0-1-sequences shown above, the diagonal method yields

     1 1 0 1 0 0 1 0 1 0 . . .

The important property of the sequence that we have now obtained is that it cannot be the same as any of the given sequences. To see this, note that the new sequence

     differs from row 1 in term number 1
     differs from row 2 in term number 2
     . . .

Since the new sequence is not equal to any of the given sequences, we have proved the following result:

For every infinite sequence of infinite 0-1-sequences, there is an infinite 0-1-sequence that does not occur among them.

In an earlier paper, in which he proved the theorem by a different method, Cantor stated this result in the following form:

The set of all 0-1-sequences is uncountable.

This theorem is the first major step in the modern theory of sets that he founded. Cantor's diagonal method is not a recursive definition. In constructing the new sequence, we do not need to know what has happened in the previous $n - 1$ steps as we proceed to the nth step. The method does, however, suggest a recursive process in which we apply the Cantor diagonal method over and over again in an attempt to list all possible 0-1-sequences. It can be shown that any such attempts are doomed to failure. In order to list all possible 0-1-sequences, a more sophisticated recursion process is required, a process that we call *transfinite recursion*. Furthermore, we also need to make use of a fundamental axiom of set theory known as the *axiom of choice*.

We mention finally that the ideas behind Cantor's diagonal method played a fundamental role in the theory of formal logic of about 1930, when Kurt Gödel (1903–1978) used the idea in order to prove his undecidability theorem (Gödel [1931]).

Chapter IV · *Optimization, Game Theory, and Economics*

MANY of the common processes of everyday life require us to coordinate a host of separate actions in order to achieve a certain goal. Some such actions may depend upon one another; it might happen that one cannot begin until certain others have been completed. Moreover, each action takes a certain amount of time; for example, an action might involve time-consuming calculations. Nowadays, we rely on systematic procedures to do all the work for us. Typical buzzwords for such procedures are network, backward calculation, and countdown.

Whenever we have found one way to accomplish a given task, we ask ourselves what other ways there may be of doing the job. The moment we gain an understanding of several possible ways in which a given task may be performed, we begin to ask ourselves how these ways be evaluated, and how the most profitable one might be selected. Such a method of selection is what we call *optimization*. For examples, we may try to buy goods of a given quality at the lowest possible price, try to obtain the highest possible remuneration for a given effort, or want to cure a disease with a minimum of undesirable side effects. During the process of optimization, we may have to deal simultaneously with completely different kinds of valuations of our choices, some of which may be easy to add and compare. For example, it is easy to add and compare sums of money; however, other kinds of valuations, like personal preferences, can be quantified only in a fairly artificial manner. For problems of this type, mathematics offers an elaborate "utility" theory whose basic beginnings date back about half a century (see, for example, Roberts [1979]).

Even if we were able to solve (at least in principle) all optimization problems of the kind that we have been describing, it may happen that different people sharing a certain situation may want to optimize their outcomes in different ways. This leads to competition. The final result would then depend upon the interplay of the strategies of several players. This *game-theoretical viewpoint* suggests a classification of such problems according to the number of interested parties involved.

One-person games are optimization problems in the classical sense.

Two-person games represent the simplest, and therefore the most fully analyzed, situation in which several players interact with one another.

However, even in the case of *n-person games* with $n \geq 3$, a variety of exact results are known that allow us to solve many n-person games. These results may apply even for very large values of n, which cause each single player in the game to "vanish in the crowd."

The purpose of this chapter is to take a brief journey through the vast field of problems that can be described in this manner. This field is divided into many subfields, and we shall give bibliographical references to these during the course of the chapter. At this stage, however, we can give the references Davis [1970] and Owen [1968] for the subject "game theory." Davis is written especially for nonmathematicians, while mathematicians will find Owen helpful.

§1   *Optimization Problems*

In this section we introduce the reader to several types of optimization problems by presenting typical examples. A standard reference for mathematicians is Collatz-Wetterling [1966].

1.1   Sorting

One kind of optimization problem is finding the largest in a given list of n numbers. We can certainly solve this problem by comparing each of the given numbers with the $n-1$ others. On the face of it, we seem to need $n(n-1)$ comparisons, but since the comparison of numbers a and b always leads to the same result as the comparison of b and a, we can at once reduce the number of necessary comparisons to $n(n-1)/2$. As a matter of fact, if all we want to do is find the largest number in the list, then we can do the job with far fewer than $n(n-1)/2$ comparisons. The reason for this is that the $n(n-1)/2$ comparisons do much more than merely find the largest number in the list. Because these comparisons compare every pair of numbers in the list, they produce a complete sorting of the list.

Let us suppose that the numbers are $z_1, \ldots, z_n$. If we start by comparing z_1 consecutively with z_2, z_3, and so on, then we have to take the following two cases into account:

Case I: The result of these $n-1$ comparisons is that $z_1 \geq z_2, z_1 \geq z_3, \ldots, z_1 \geq z_n$. In this case, we know after the $n-1$ comparisons that z_1 is the largest number in the list.

Case II: The result $z_1 < z_k$ occurs at least once. We define k to be the first index for which this happens. From now on we know that the numbers $z_1, \ldots, z_{k-1}$ are not candidates for the position of largest number. It took us $k-1$ comparisons to learn this. We now continue by comparing z_k with $z_{k+1}, \ldots$, and once again we encounter Case I or Case II.

It is evident that this procedure, or *algorithm* as we call it, will yield the maximum of the numbers $z_1, \ldots, z_n$ after $n-1$ comparisons. Note that the job cannot be done with fewer than n comparisons because in this case there would be at least two numbers that are never compared in the procedure. Therefore, our procedure is *optimal*. For a systematic treatment of "sorting and searching" see Knuth [1968] vol. 3.

1.2   A Simple Case of Linear Optimization

In recent times, linear optimization has become a common procedure in many business offices. In the simple case we are considering, the problem is to find two numbers x and y such that a given set of linear inequalities holds, and such that a given linear expression in x and y is a maximum (or a minimum). Note that linear expressions in x and y are polynomials in x and y that contain x and y to the first power only. No terms such as x^2, y^2, xy, x^3 are allowed. A typical linear inequality in x and y is $3x - 2y \leq 4$. We shall now consider an example (see Franklin [1980]).

A bank has 100 million dollars to invest. It invests x million dollars in the form of loans to individuals at an interest rate of 10 percent, and invests y million dollars in shares that pay dividends at a rate of 5 percent. We need to find values of x and y that will maximize the bank's revenue, which is $0.10x + 0.05y$ million dollars. The bank must, of course, obey the following three obvious constraints:

$$x \geq 0$$
$$y \geq 0$$
$$x + y \leq 100,$$

but there are also other constraints that have to be taken into account. By law, a bank must keep at least 25 percent of its investments in reserve. Therefore $y \geq 0.25(x + y)$, and so $x \leq 3y$. Furthermore, the bank expects its clients to ask for loans totaling at least 30 million dollars, so $x \geq 30$. We note that these constraints amount to five linear inequalities in x and y. We illustrate each of these by showing it as a shaded region of points (x,y) in Figure IV.1.1. The points (x,y) that satisfy all five of these constraints

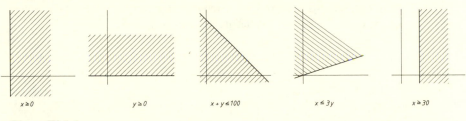

Figure IV.1.1

form the polygon G that is obtained by intersecting the five shaded regions. This polygon G is the triangle ABC shown in Figure IV.1.2. In this figure, the lines of the form $0.10x + 0.05y = $ constant are as shown in Figure IV.1.3. Among these lines, we need to choose one that intersects with the triangle G such that the constant is as large as possible.

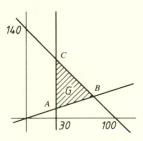

Figure IV.1.2

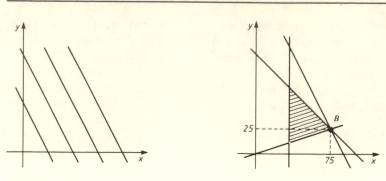

Figure IV.1.3   **Figure IV.1.4**

We therefore need a line that is as far to the right as possible. This line passes through the point B shown in Figure IV.1.4. From this we deduce that the revenue is maximal when the bank puts 75 million dollars into loans and places the remaining 25 million dollars in stocks. The maximal revenue in millions of dollars is $(0.10)(75) + (0.05)(25) = 8.75$.

1.3   An Example of Quadratic Optimization

Among all rectangles with a given circumference C, which one has the largest area F? If x, y are the two sides of the rectangle, we certainly have

$$F = x \cdot y$$
$$C = 2x + 2y.$$

The second equality yields

$$y = \frac{1}{2}\left(C - 2x\right),$$

and thus F can be written as a function of x alone:

$$F = F(x) = x\frac{1}{2}\left(C - 2x\right).$$

Now we calculate

$$F(x) = -x^2 + \frac{C}{2}x$$

$$= -x^2 + 2\frac{C}{4}x$$

$$= -x^2 + 2\frac{C}{4}x - \left(\frac{C}{4}\right)^2 + \left(\frac{C}{4}\right)^2$$

$$= -\left(x - \frac{C}{4}\right)^2 + \left(\frac{C}{4}\right)^2$$

$$= \frac{C^2}{16} - \left(x - \frac{C}{4}\right)^2.$$

Surely this expression is maximal if the subtracted squared bracket $\left(x - \frac{C}{4}\right)^2$ takes on its minimal value, which happens precisely for $x = \frac{C}{4}$; this in turn implies $y = \frac{C}{4} = x$:

the rectangle of maximal area is a *square*. What you have seen here is one of the simplest cases of a so-called *isoperimetric problem*. A famous result in this field is the following: among all plane figures with a given circumference, the circle has the largest area. Its proof is beyond the scope of this book.

1.4   An Example of Cubic Optimization

Suppose that we have a cardboard square of side 1 unit, and we cut out a smaller square of side x from each of the four corners of this square. We now make a box (without a top) by folding the cardboard as shown in Figure IV.1.5. If $f(x)$ is the volume of this box, then we have

$$f(x) = (1 - 2x)(1 - 2x)x = 4x^3 - 4x^2 + x.$$

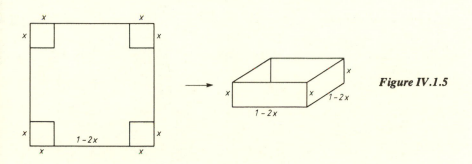

Figure IV.1.5

Our problem is to choose the value of x that will make the volume of this box as large as possible. A glance at Figure IV.1.5 tells us that $0 \le x \le 1/2$. Furthermore, we see that

when $x = 0$, we have $f(x) = 0$,
when $0 < x < 1/2$, we have $f(x) > 0$,
when $x = 1/2$, we have f(x) = 0.

The graph of this function f is shown in Figure IV.1.6.

Since the maximum value of $f(x)$ does not occur when x is 0 or 1/2, the value of x at

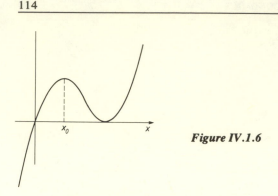

Figure IV.1.6

which $f(x)$ is maximum must lie between 0 and 1/2. To find this value x_0, we shall use a little elementary calculus.

We see that for each x,

$$f'(x) = 12y^2 - 8x + 1 = (2x - 1)(6x - 1),$$

and since the only value of x between 0 and 1/2 at which $f'(x) = 0$ is 1/6, we conclude that the maximum volume of the box will be attained when $x = 1/6$. Thus our maximal box is the one shown in Figure IV.1.7, and the maximum volume is

$$\frac{2}{3} \cdot \frac{2}{3} \cdot \frac{1}{6} = \frac{2}{27}.$$

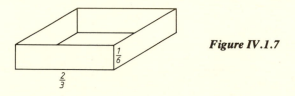

Figure IV.1.7

One can easily see that other values of x will give a smaller volume. For example, when $x = 1/4$, the volume is only $1/16 = 2/32 < 2/27$.

We call this problem a problem of cubic optimization because it involved the maximizing of a cubic polynomial.

1.5   Network Optimization

Imagine a town whose streets allow interurban traffic to pass through as shown in Figure IV.1.8. Suppose that the roads on each inside edge have the capacity to support one truck every five seconds. Our problem is to decide how the traffic should be regulated in order to allow a maximum number of trucks to pass through in a given amount of time (five seconds, for example). In a simple situation like this it is possible to guess the

Figure IV.1.8

optimal solution, and it is also possible to list all the possibilities. However, as the problem becomes increasingly more complicated, we need to find a systematic, step-by-step approach, an *algorithm*, that will provide the optimal solution. We shall see such an algorithm in the next section.

§2   *Optimal Flows in Networks*

We shall not give the formal definition of a network in this section. Instead, we shall look at a network informally in the following way.

(1) A network N is a collection of *arrows* (or *edges*) that connect *points* (or *vertices*).

(2) There is a unique point Q (called the *source* of the network) at which no arrow ends but at least one arrow begins. There is also a unique point S (called the *sink* of the network) at which no arrow starts but at least one arrow ends. At every other point of the network, at least one arrow starts and at least one arrow ends.

(3) A sequence of arrows (as depicted in Figure IV.2.1) is said to be a *path* if the tip of each arrow in the sequence coincides with the tail of the next arrow. We shall not consider networks that contain circular paths (*cycles*).

It is easy to see that every path in a network that starts at the source Q can be extended into a path that ends at the sink S. To see this, we need only observe that any path that does not end at S can be extended farther, and since the network contains no cycles, extension of the path must eventually yield a path that ends at S. We deduce that a network always has at least one (and possibly many) paths that run from the source to the sink.

A *flow* in a network N is an assignment of a nonnegative number $f(e)$ to each of the arrows e in N. We may think of $f(e)$ as being the amount of flow in the arrow e. A flow is required to satisfy the condition that for every point P in the network, other than Q and S, the total flow into P = the total flow out of P. This property of a flow is an abstract version of Kirchhoff's law, and can be thought of as meaning that there is no creation or destruction of matter in the network. Using this law it is easy to show that the total flow out of Q = the total flow into S. This number is called the *strength* of the given flow f, and is written as $\|f\|$.

A *capacity distribution* over a given network N is also an assignment of nonnegative numbers $c(e)$ to each of the arrows e in N. If c is a capacity distribution over a network N, and f is a flow in N, then we shall say that the flow f *respects* the capacity distribution c if we have $f(e) \leq c(e)$ for every arrow e in the network. We write this condition more

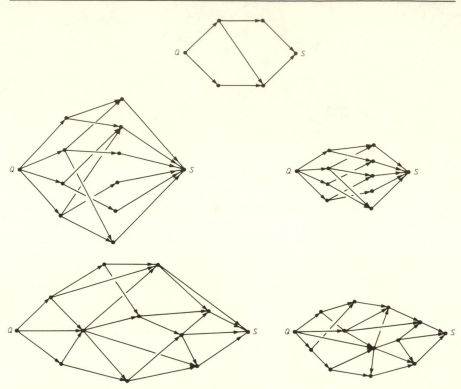

Figure IV.2.1

simply as $f \le c$. In the event that $f(e) = c(e)$ for a given arrow e, the arrow is said to be *full*. If $f(e) = 0$, the arrow e is said to be *dead*, and if $f(e) > 0$ the arrow e is said to be *busy*.

We can now state the *flow optimization problem* for a given capacity distribution c over a given network N.

> *Flow Optimization Problem:* Given a capacity distribution c over a network N, find a flow f in N that respects c, and that has largest possible strength $\|f\|$.

There will usually be several such flows f in a given network. In the example that appeared at the end of Section 1, it was easy to find a maximal flow by just guessing. The following *labeling algorithm* will show how to arrive at a maximal flow in any given network, no matter how complicated it may be, and no matter how its capacity distribution c has been chosen, provided that all of the numbers $c(e)$ are integers. The basic idea is as follows:

> (1) We begin with any flow $f_1 \le c$; for example, with the "zero flow" that assigns the value $f_1(e) = 0$ to every arrow e in N.
>
> (2) We replace f_1 by a new flow $f_2 \le c$ such that $\|f_2\| > \|f_1\|$. We call this step an *augmentation step*.

(3) We repeat step (2) until no further augmentation steps are possible. At this stage we have constructed a finite sequence $f_1, f_2, \ldots, f_n$ such that $\|f_1\| < \|f_2\| < \ldots < \|f_n\|$ and

$$\|f_n\| = \max_{f \leq c} \|f\| .$$

The flow $f = f_n$ solves the problem.

From now on, we shall restrict ourselves to networks in which the values of the numbers $c(e)$ and $f(e)$ are always integers. We call such flows *integer flows*. We shall design the augmentation step in such a way that in increases the strength of the flow by 1 each time, until a maximal flow is obtained. Since all the numbers $c(e)$ are finite integers, the process must end after a finite number of steps.

There is one obvious idea how to augment a given flow $f_1 \leq c$. We might find a path from Q to S such that no arrow in this path is full, and then augment $f_1(e)$ by 1 every arrow e in this path. This process yields a new flow f_2 such that $f_2 \leq c$ and $\|f_2\| = \|f_1\| + 1$. Unfortunately, this simple method does not always lead to a maximal flow, as can be seen from the example given in Section 1. In that example, we may define $c(e) = 1$ for every e in N and then define f_1 as shown in Figure IV.2.2. Since every path from A to S in this network has at least one full arrow, the preceding method cannot be applied even though there is a flow f satisfying $\|f\| > \|f_1\|$, namely the one shown in Figure IV.2.3.

Figure IV.2.2   **Figure IV.2.3**

We shall therefore improve the preceding method by allowing some of the values $f_1(e)$ to decrease in the augmentation step while others increase, as does $\|f\|$. The method we are about to describe is known as the *labeling algorithm* of Ford-Fulkerson [1956]. In this algorithm, the augmentation step consists of a procedure of assigning labels to certain vertices of the given network and then using these labels to modify the given flow f_1. The labeling procedure is carried out according to the following rules:

(1) We start by giving a label to the source Q.
(2) We continue labeling according to the following routines:
    (a) Forward labeling: If we have already given a label to the tail of an arrow that is not full, then we give a label to its tip, too.
    (b) Backward labeling: If we have already given a label to the tip of a busy arrow, then we give a label to its tail, too.
    The labeling procedure terminates as soon as we have no further opportunities to apply these routines.

In the event that this labeling procedure penetrates as far as the sink S, then we reconstruct the succession of steps (a) and (b) that led us to label S, and we modify the given flow f_1 as follows:

> Replace $f_1(e)$ by $f_1(e) + 1$ if e occurred during an application of step (a) (forward labeling).

> Replace $f_1(e)$ by $f_1(e) - 1$ if e occurred during an application of step (b) (backward labeling).

We now make three observations:

> The change from $f_1(e)$ to $f_1(e) + 1$ was made only in arrows e that were not full. Therefore, the new flow still respects the capacity constraints.

> The change from $f_1(e)$ to $f_1(e) - 1$ was made only in busy arrows. Therefore, the values of the new flow are still nonnegative.

> Kirchhoff's law still holds for the new flow. We leave the proof of this fact to the reader.

> If we call the new flow f_2, then $\|f_2\| = \|f_1\| + 1$: one unit more is going out of the source Q.

We now repeat this procedure of labeling and augmenting again and again, until it can no longer be carried out. The procedure will certainly end after a finite number of steps because the strength of the flow can be increased by 1 only a finite number of times (as long as the given capacity constraints have to be respected). Now, suppose that f is the flow that results from this finite number of steps.

The important property of f is that if the labeling procedure is now applied to f, then the sink S is *not* labeled, and therefore the method just described can no longer be used to increase the strength of the flow. We nevertheless apply the labeling procedure to f until neither (a) nor (b) can be carried out, and note that the point S has not been labeled.

Consider all arrows that now bear a label at their tail but not at their tip. Any path from Q to S begins with the labeled vertex Q and ends with the unlabeled vertex S, and hence contains at least one arrow of the class just defined. Therefore, if we remove these particular arrows, it will no longer be possible to have a nonzero flow. These arrows therefore form what is known as a *cut*. The arrows in this cut are all full—otherwise, the routine (a) could still have been applied, contrary to our assumption. A simple calculation now shows that $\|f\|$ is the sum of all the capacities $c(k)$ where k is an arrow that belongs to the cut. We shall call the cut C_0 and define $c(C_0)$ to be the sum of all the capacities $c(k)$ where k belongs to C_0. We now have

$$\|f\| = c(C_0).$$

Now, given *any* flow f and *any* cut C, the inequality $f \leq c$ easily yields

$$\|f\| \leq c(C).$$

Keeping f fixed and varying the cut C, we obtain

$$\|f\| \leq \min_{C} c(C)$$

for any flow f that satisfies $f \leq c$. If we now let f vary subject to the requirement $f \leq c$, we obtain

$$\max_{f \leq c} \|f\| \leq \min_{C} c(C).$$

Our previous equality $\|f\| = c(C_0)$ now tells us that the latter inequality is an equation. In other words,

$$\max_{f \leq c} \|f\| \leq \min_{C} c(C).$$

This statement is the well-known Ford-Fulkerson *max-flow-min-cut theorem*. Since the augmentation step in this algorithm carries integral flows into integral flows, we obtain the following additional information: if all values $c(e)$ are integers, then there is a maximal flow f for which all values $f(e)$ are integers.

We have given the proof of Ford-Fulkerson's theorem in some detail because it provides insight into *linear optimization*, which is a field of enormous practical importance nowadays, and also because both the theorem and the proof are easy to visualize. As an example of an application of this theorem within mathematics, we now give a network proof of the marriage theorem.

Network Proof of the Marriage Theorem. We recall from Section 4 of Chapter III that we have a system of friendships among a finite number g of gentlemen and a finite number d of ladies (see Figure IV.2.4.). We want to provide a husband to each of the ladies in such a way that no lady marries a gentlemen who is not one of her friends. The marriage theorem says that this can be done if and only if the *party condition* holds. This condition says that any k ladies will always have at least k friends between them. We shall now assume that the party condition is satisfied and we shall marry off the ladies by constructing a network out of the given friendship graph. We artificially add in a source Q and a sink S, and link the ladies to the source and the gentlemen to the sink (see Figure IV.2.5).

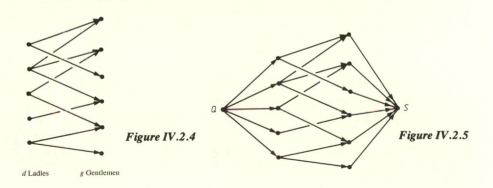

Figure IV.2.4

Figure IV.2.5

d Ladies g Gentlemen

Next we assign capacities: We assign $c(e) = 1$ to each of the new arrows e, left and right, and assign some very large integer values $c(e)$ to each of the original friendship arrows in the middle.

We now show that the arrows starting from Q form a minimal cut C_0. To obtain a contradiction, suppose that there is another cut C such that $c(C) < c(C_0) = d$. The cut C certainly cannot contain any original friendship arrows because if it did, the large values $c(e)$ of such arrows e would make it impossible to have $c(C) < c(C_0)$. Therefore C consists only of some new arrows to the left and some new arrows to the right. Since $c(e) = 1$ for every one of these arrows, the inequality $c(C) < c(C_0)$ simply means that there are fewer arrows in C than there are in C_0. Now C_0 has exactly d arrows, where d is the number of ladies. If we pass from C_0 to C by removing, for example, a left arrows and adding at most $a-1$ right arrows, we single out a group of a ladies whose incoming arrows do not belong to C. Since C is a cut, the friendship arrows starting from these a ladies must all end with the at most $a-1$ gentlemen whose outgoing arrows belong to C. Since this statement violates the party condition, we have arrived at the desired contradiction. Thus the minimality of $c(C_0)$ is established. Ford-Fulkerson's theorem now gives us a flow f (of love)

$$\|f\| = c(C_0) = d = \text{the number of ladies,}$$

and for which all values $f(e)$ are integers. Since no lady can store the unit of love she receives, she must give it away to someone; and since all the values $f(e)$ are integers, the lady gives it away to precisely one gentleman. Furthermore, no gentleman is assigned to more than one lady because the capacity $c(e)$ of the arrow e linking him to the sink S is 1: he can't accept more than one unit. We thus arrive at a monogamous marriage system, and have once again proved the marriage theorem.

This proof of the marriage theorem illustrates a method of proof frequently used in mathematics: we restate a problem originally from theory A anew in terms of theory B, and in this way the methods of theory B become available for the solution of the problem.

Flow optimization is a linear problem. All quantities occur to the first power only (not squared, and so on). The labeling algorithm that we have described is a common tool of linear optimization. This algorithm is closely related to the most important tool of the theory of linear optimization: the *simplex algorithm* (Dantzig [1951]). Anyone who could earn royalties from each application of the simplex algorithm would drown in money.

§3   *Ordered Fields*

Order relationships such as ''greater than'' or ''less than'' played only an occasional role in the material of Chapter II, but these relationships constitute the main theme of the present chapter. We shall therefore discuss some of the basic facts about order relations; in particular, when we have an order in an algebraic system (such as a field) we shall be concerned with possible interplay between the order and the arithmetical operations of the system. The basic concept of this section is that of an *ordered field*.

3.1 Orders and Partial Orders

Roughly speaking, an order $\leq$ in a set X is a way of saying for any two elements x and y of X whether a statement $x \leq y$ (which we read as "x is less than or equal to y") is true or false. More precisely, we obtain an order $\leq$ in a set X by partitioning the set of all ordered pairs (x,y) of elements of X into two classes. When a pair (x,y) belongs to the first class we write $x \leq y$, and when (x,y) belongs to the second class we say that the condition $x \leq y$ is false. Furthermore, to be called an order, a relation $\leq$ must obey the following three laws:

(1) $x \leq x$ for every x in X. This law is called the *reflexivity law*.
(2) Whenever $x \leq y$ and $y \leq z$, then $x \leq z$. This law is called the *transitivity law*.
(3) Given any two elements x and y in the set X, either $x \leq y$, or $y \leq x$. We include the possibility that both of these conditions may hold. This law is called the *law of comparability*.

In short, a binary relation $\leq$ in a set X is said to be an order in X if $\leq$ is reflexive and transitive, and satisfies the comparability law.

Given two elements x and y of X, the condition $x \geq y$ means simply that $y \leq x$.

There is a fourth law, called the *law of strictness*, that we shall usually require to hold true. This law says that if x and y are elements of the set X and both of the conditions $x \leq y$ and $y \leq x$ hold, that we have $x = y$.

When $x \leq y$ and $x \neq y$, we write $x < y$. The relation thus defined is still transitive, but it is not reflexive. Using the relation $<$ we can restate the law of comparability as follows:

For any two elements x and y of the set X, one and only one of the conditions

$$x < y, \qquad x = y, \qquad y < x$$

will hold.

Note, for example, that if we had $x < y$ and $y < x$, then the law of strictness would be violated.

The classic example of an order is the usual order in the real line $\mathbb{R}$. In this case, the condition $x \leq y$ can be represented geometrically by saying that x "lies to the left" of y. Don't forget that the real line is a manmade mathematical object and cannot be found in nature. In particular, the order $\leq$ in $\mathbb{R}$ has been chosen by mathematicians. They could as well have chosen it in a different way, but this particular choice has the properties that we desire. In general, we may consider a variety of orders in a given set X. We may for example use an order to express degrees of desirability; such orders are known as "preference orders." If a set X has n elements, then there are $n! = n \cdot (n-1) \cdot \ldots \cdot 2 \cdot 1$ possible orders in X. For example, if $n = 3$ then there are six possible orders of the set. If the set contains elements a, b, and c, then the orders in X are as shown in the columns below.

$$a \; a \; b \; b \; c \; c$$
$$b \; c \; a \; c \; a \; b$$
$$c \; b \; c \; a \; b \; a$$

A relation in a set that satisfies the conditions of reflexivity, transitivity, and strictness, but which may not satisfy the comparability law, is said to be a *partial order* in the set. Note that every order is a partial order. A partial order fails to be an order if and only if there exist two elements x and y in the set X such that neither of the conditions $x \leq y$ and $y \leq x$ are true. Such elements are said to be *noncomparable*. An example of a partial order that is not an order is the partial order in the Euclidean plane $\mathbb{R}^2$ defined by

$A \leq B$ if and only if A lies to the lower left of B.

More precisely, if A and B are the points with coordinates (x_1, y_1) and (x_2, y_2), respectively, then the condition $A \leq B$ means that $x_1 \leq x_2$ and $y_1 \leq y_2$. Two points that are noncomparable are depicted in Figure IV.3.1.

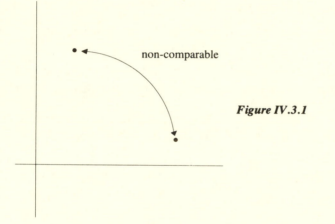

non-comparable

Figure IV.3.1

In Figure IV.3.2 we show some graphical representation of partial orderings in finite sets. The condition $x \leq y$ is visually defined by saying that we can walk upwards from x to y along edges in the diagram. It is easy to see that the reflexivity, transitivity, and strictness laws all hold.

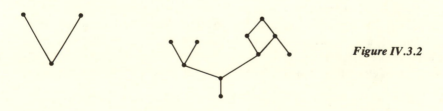

Figure IV.3.2

3.2   Ordered Fields

As we have seen, a set may be ordered or partially ordered in many different ways. However, if the given set is already a field (see Chapter II, Section 2), then we are particularly interested in those orders which are in natural harmony with the arithmetic operations of addition and multiplication in the field. The kind of harmony that we desire is described in the following definition.

Definition. Suppose that K is a field with addition $+$, with multiplication $\cdot$, and with additive and multiplicative identities 0 and 1, respectively. Suppose that $\leq$ is an ordering of the set K. We say that K together with the ordering $\leq$ is an *ordered field* if the following two compatibility conditions are satisifed:

(1) Compatibility of $\leq$ with addition:
Whenever $a \leq b$, we have $a + c \leq b + c$ for every element c in K.
More briefly we write
$a \leq b \Rightarrow a + c \leq b + c$ for every c in K.

(2) Compatibility of $\leq$ with multiplication:
$a \leq b$ and $c \geq 0 \Rightarrow a \cdot c \leq b \cdot c$.

A classic example of an ordered field is the field $\mathbb{R}$ of all real numbers with its usual ordering. The subfield $\mathbb{Q}$ of rational numbers is another natural example. For all of our present applications, the field $\mathbb{R}$ is sufficient. There are, however, many other ordered fields. The following are some properties that are logical consequences of (1) and (2) and hence valid in every ordered field:

(3) $a \leq a'$ and $b \leq b' \Rightarrow a + b \leq a' + b'$.

We may deduce this fact by applying rule (1) twice. Assuming that $a \leq a'$ and $b \leq b'$, we have $a + b \leq a' + b$ and $a' + b \leq a' + b'$, and the result $a + b \leq a' + b'$ now follows from transitivity.

(4) $a \leq a'$ and $b < b' \Rightarrow a + b < a' + b'$.

From (1) we see that $a + b \leq a + b'$. However, we cannot have $a + b = a + b'$, because then we would have $b = b'$. Therefore we have $a + b < a + b'$. Since $a + b' \leq a' + b'$, condition (4) follows from transitivity.

(5) $0 \leq a \leq a'$ and $0 \leq b \leq b' \Rightarrow 0 \leq ab \leq a'b'$.

Using the condition $0 \leq a \leq a'$ and (2) we deduce that $0 \cdot b \leq a \cdot b \leq a' \cdot b$, and therefore $0 \leq ab \leq a'b$. Using (2) again we see that $a'b \leq a'b'$, and condition (5) now follows from transitivity.

(6) $0 < a < a'$ and $0 < b < b' \Rightarrow 0 < ab < a'b'$.

From (5) we know that $0 \leq ab \leq a'b'$. The condition $0 = ab$ is impossible because it implies that either $a = 0$ or $b = 0$, and hence $0 < ab$ follows. In a similar way we may observe that $ab < a'b$ and $a'b < a'b'$, and the fact that $ab < a'b'$ now follows from transitivity.

(7) $a \leq b \Rightarrow -a \geq -b$.

We observe that
$$a \leq b \Rightarrow a + (-a) \leq b + (-a) \Rightarrow 0 \leq b - a$$
$$0 + (-b) \leq b - a + (-b) \Rightarrow -b \leq -a \Rightarrow -a \geq -b.$$

(8) *Whenever $a \neq 0$ we have $a^2 > 0$*, i.e., *squares of nonzero elements are always positive.*

If $a > 0$ then it follows from (6) that $a^2 > 0$. If $a < 0$, then since $-a > 0$, it again follows from (6) that $a^2 = (-a)^2 > 0$.

(9) $1 > 0$.

Clear from (8) since $1 = 1^2 > 0$.

(10) $a \leq b$ and $c < 0 \Rightarrow ac \geq bc$.

(11) $a < b$ and $c > 0 \Rightarrow ac < bc$.

From (2) we deduce that $ac \leq bc$, and because $a \neq b$ we cannot have $ac = bc$.

(12) $a > 0 \Rightarrow 1/a > 0$, and $a < 0 \Rightarrow 1/a < 0$.

If we had $a > 0$ and $1/a \leq 0$, then we would have $1 = a(1/a) \leq 0$, contradicting (9). Therefore the first statement holds, and we may deduce the second statement in a similar fashion.

(13) $0 < a < b \Rightarrow \dfrac{1}{a} > \dfrac{1}{b} \Rightarrow \dfrac{c}{a} > \dfrac{c}{b}$ *for all $c > 0$.*

This law says that if we enlarge the denominator of a positive fraction, the fraction becomes smaller. To prove this law we observe first, using (6), that $ab > 0$. From (12) we deduce that $1/(ab) > 0$. Therefore, since $a < b$ we have

$$a\left(\frac{1}{ab}\right) < b\left(\frac{1}{ab}\right),$$

which gives $1/b < 1/a$. The last inequality is now obtained upon multiplying by c.

Given an element a of an ordered field K, there are two possibilities: either $a \geq 0$, or $a < 0$, in which case $-a > 0$. We define the *absolute value* or *modulus* $|a|$ of the element a by

$$|a| = \begin{cases} a & \text{if } a \geq 0 \\ -a & \text{if } a < 0. \end{cases}$$

We see that $|a| \geq 0$ in all cases, and that $|-a| = |a|$. Furthermore,

$$-|a| \leq a \leq |a|$$

and

$$a \geq b \text{ and } a \geq -b \qquad a \geq |b|.$$

We may write this last statement in the form

$$|b| \leq a \Leftrightarrow -a \leq b \leq a.$$

(14) The so-called *triangle inequality* for the absolute value says that for all elements a and b we have $|a + b| \leq |a| + |b|$.

To prove this law we use the facts that $-|a| \leq a \leq |a|$ and $-|b| \leq b \leq |b|$ and rule (4) to deduce that

$$-(|a| + |b|) \leq a + b \leq |a| + |b|,$$

and from this fact the triangle law follows at once.

(15) For all elements a and b we have

$$\left||a| - |b|\right| \leq |a - b|.$$

To prove this law we note first that

$$|a| = |a - b + b| \leq |a - b| + |b|,$$

and therefore

$$|a| - |b| \leq |a - b|.$$

Similarly we have

$$|b| - |a| \leq |b - a| = |-(a - b)| = |a - b|.$$

Since $\left||a| - |b|\right|$ is either $|a| - |b|$ or $|b| - |a|$, the desired result has been proved.

(16) For all elements a, b we have $|a \cdot b| = |a| \cdot |b|$.

To prove this result, we separate it into four cases:

> *Case I*: $a \geq 0$ and $b \geq 0$. In this case we have $ab \geq 0$, and so the desired equation simply says that $ab = ab$.
>
> *Case II*: $a < 0$ and $b < 0$. In this case we use Case I and deduce that $|ab| = |(-a) \cdot (-b)| = |(-a)| \cdot |(-b)| = |a| \cdot |b|$.
>
> *Case III*: $a < 0$ and $b \geq 0$. In this case we use Case I and deduce that $|ab| = |-ab| = |(-a)b| = |-a| \cdot |b| = = |a| \cdot |b|$.
>
> *Case IV*: $a \geq 0$ and $b < 0$. The proof in this case is analogous to the proof of Case III.

One of the most important statements that one might make about the relationship between the order and the arithmetical operations in an ordered field involves the role of the set of natural numbers. We mention first that to every natural number n, we can associate an element $n \times 1$ in the given field that is made by adding the element 1 to itself n times. The statement that follows does not hold in every ordered field, and is known as the *Archimedean property*.

> *Archimedean property*. Given any element $a > 0$, there is a natural number n such that $n \times 1 > a$.

An element a of an ordered field is sometimes called *infinitely large* if $a \geq n \times 1$ for every natural number n. Using this notion we can say that a given ordered field has the Archimedean property if it has no infinitely large elements.

Fields with the Archimedean property are called *Archimedean fields*. (There are examples of non-Archimedean fields, but we will not dwell on them here.) $\mathbb{R}$ is an Archimedean field; this is one of its most important properties. Note that if a field is Archimedean and $a > 0$, then since $1/a > 0$, we can find a natural number n such that $n \times 1 > 1/a$. This natural number n has the property that

$$\frac{1}{n \times 1} < a.$$

We conclude this subsection with the following remark:

Ordered fields are always infinite.

We can see this by noting that since $1 > 0$, we have $2 \times 1 > 1$, and, continuing in this manner, we can see that the field has an infinite number of elements of the form $n \times 1$, where n is a natural number. From this fact we deduce that finite fields such as $GF(2)$ and $GF(3)$ cannot be ordered.

3.3 Arithmetical Means and Weighted Means

In this subsection we increase our understanding of the ordered field $\mathbb{R}$ of real numbers by looking at means (averages). The *arithmetic mean* of a finite sequence $a_1, a_2, \ldots, a_n$ is defined to be the number

$$a = \frac{1}{n}\left(a_1 + \ldots + a_n\right) = \frac{1}{n}\sum_{k=1}^{n} a_k.$$

The arithmetic mean of a set of data tells us where, roughly, the data can be expected to lie.

Many of the results of this subsection will also hold for a somewhat more general mean that we call the *weighted mean*. Given nonnegative numbers $g_1, \ldots, g_n$ satisfying the condition $g_1 + \ldots + g_n = 1$, the weighted mean of the sequence $a_1, a_2, \ldots, a_n$ with *weights* $g_1, \ldots, g_n$ is defined to be the sum

$$g_1 a_1 + \ldots + g_n a_n = \sum_{k=1}^{n} g_k a_k.$$

The arithmetical mean is the special case of the weighted mean that results when

$$g_1 = \ldots = g_n = \frac{1}{n}.$$

Weighted means can play an important role in many familiar situations. Suppose, for example, that the oral examination given in a college course is only half as difficult as the written examination. It would be reasonable to calculate the overall grade a of a student in the course from the oral grade a_1 and the written grade a_2 by using the formula

$$a = \frac{1}{3}a_1 + \frac{2}{3}a_2,$$

rather than by using the arithmetic mean

$$a = \frac{1}{2}a_1 + \frac{1}{2}a_2.$$

In the weighted mean the written grade is twice as important as the oral one. Even political decisions need to take account of weights, as can be seen from the following example.

Student Admission and Sex Bias at Berkeley in 1973. A total of 8442 men and 4321 women applied for admission at the University of California, Berkeley, for the fall quarter, 1973. The university admitted 44 percent of the male applicants and only 35 percent of the female applicants. It was, of course, immediately accused of sex bias against women. However, when an investigating team looked into the details, they found the following figures showing the numbers of applicants for the six most sought-after programs of study (see Figure IV.3.3).

Figure IV.3.3

Subject	Males		Females	
	Applicants	Admitted (percent)	Applicants	Admitted (percent)
A	825	62	108	82
B	560	63	25	68
C	325	37	593	34
D	417	33	375	35
E	191	28	393	24
F	373	6	341	7
total	2691	44%	1835	30%

Except, perhaps, for subject A, the admission percentages for men and for women in each subject were roughly the same. The alleged "sex bias" was the result of different preferences among the men and women. The men had preferred A and B, which had fairly high admission quotas, while the women had concentrated on the low-quota programs C, D, E, and F, and were therefore at a disadvantage when the figures were looked at globally. It seems more fair to use the numbers of applicants for the various programs as a measure for the weights in a weighted average (see Figure IV.3.4). Using this weighted average we obtain the following figures:

Men: $g_A\ 62 + g_B\ 63 + g_C\ 37 + g_D\ 33 + g_E\ 28 + g_F\ 6 = 39\%$

Women: $g_A\ 82 + g_B\ 68 + g_C\ 34 + g_D\ 35 + g_E\ 24 + g_F\ 7 = 43\%$

Figure IV.3.4

Subject	Applications (total)	Weight
A	933	$g_A = \dfrac{933}{4526}$
B	585	$g_B = \dfrac{585}{4526}$
C	918	$g_C = \dfrac{918}{4526}$
D	792	$g_D = \dfrac{792}{4526}$
E	584	$g_E = \dfrac{584}{4526}$
F	714	$g_F = \dfrac{714}{4526}$
total	4526	

Now it was the men's turn to complain about "sex bias."

One might ask why it is reasonable to calculate an average $(a_1 + \ldots + a_n)/n$ or a weighted average $g_1 a_1 + \ldots + g_n a_n$ to give a rough characterization of the position of a "cloud" of data. Carl Friedrich Gauss (1777–1855) put the problem as follows:

> Given the data $a_1, \ldots, a_n$ and the weights $g_1, \ldots, g_n$ (subject to the conditions we have stated), which number a minimizes the weighted mean
>
> $$(a_1 - a)^2 g_1 + \ldots + (a_n - a)^2 g_n$$
>
> of the "squared differences" from a? The answer is
>
> $$a = \tilde{a} = g_1 a_1 + \ldots + g_n a_n.$$

Proof: Given any number a, we shall compare the numbers

(1) $(a_1 - a)^2 g_1 + \ldots + (a_n - a)^2 g_n$

and

(2) $(a_1 - \tilde{a})^2 g_1 + \ldots + (a_n - \tilde{a})^2 g_n.$

The expression (1) may be written as

$$(a_1 - \tilde{a} + (\tilde{a} - a))^2 g_1 + \ldots + (a_n - \tilde{a} + (\tilde{a} - a))^2 g_n$$
$$= (a_1 - \tilde{a})^2 g_1 + 2(a_1 - \tilde{a})(\tilde{a} - a)g_1 + (\tilde{a} - a)^2 g_1$$
$$+ \ldots$$
$$+ (a_n - \tilde{a})^2 g_n + 2(a_n - \tilde{a})(\tilde{a} - a)g_n + (\tilde{a} - a)^2 g_n.$$

Now, using the fact that the numbers $g_1, \ldots, g_n$ are nonnegative and that $g_1 + \ldots + g_n = 1$, we can simplify this expression. We obtain

$$(a_1 - \bar{a})^2 g_n + \ldots + (a_n - \bar{a})^2 g_n$$
$$+ 2(a_1 g_1 + \ldots + a_n g_n - \bar{a} g_1 - \ldots - \bar{a} g_n)(\bar{a} - a)$$
$$+ (\bar{a} - a)^2.$$

This yields

$$(a_1 - \bar{a})^2 g_1 + \ldots + (a_n - \bar{a})^2 g_n$$
$$+ 2(\bar{a} - \bar{a})(\bar{a} - a) + (\bar{a} - a)^2$$
$$= (a_1 - \bar{a})^2 g_1 + \ldots + (a_n - \bar{a})^2 g_n + (\bar{a} - a)^2.$$

We have thereby shown that expression (1) exceeds expression (2) by the nonnegative number $(\bar{a} - a)^2$, and therefore expression (2) is the smallest number of this form. This characterization of the weighted mean is called the *least squares method* (Gauss, around 1800).

We shall now look at a few properties of weighted averages (and hence of ordinary averages) that follow from the laws (1)–(16) that we established previously. These properties are roughly those we need for everyday applications of this material.

(a) An increase of all the numbers $a_1, a_2, \ldots, a_n$ causes an increase in their weighted average. This property follows from the fact that the numbers $g_1, \ldots, g_n$ are all nonnegative. If $a'_i \geq a_i$ for each i, then $a'_i g_i \geq a_i g_i$ for each i, and hence

$$\sum_{i=1}^{n} a'_i g_i \geq \sum_{i=1}^{n} a_i g_i.$$

(b) If we increase one number a_j and decrease another number a_k by the same amount, and if $g_j > g_k$, then the weighted average is increased. In fact, if a_j is increased by $\epsilon > 0$ and a_k is decreased by ϵ, then the weighted average is increased by $\epsilon \cdot (g_j - g_k)$.

(c) If all of the numbers $a_1, a_2, \ldots, a_n$ are nonnegative and $\bar{a} \leq \epsilon^2$, and if g is the sum of all the numbers g_j for which $a_j > \epsilon$, then $g \leq \epsilon$. To prove this property, we observe that the sum of the terms $a_j g_j$ for which $a_j > \epsilon$ is at least ϵg, but cannot exceed $\bar{a}$. Therefore $\epsilon g \leq \epsilon^2$, and so $g \leq \epsilon$.

If we are working with ordinary averaging, then property (c) tells us that if $\bar{a} \leq \epsilon^2$, then the percentage of those numbers j for which $a_j > \epsilon$ is less than 100ϵ. For example, if $\bar{a} < 1/10000$, then fewer than 1 percent of all the numbers a_j can be greater than $1/100$.

In addition to the arithmetic mean

$$\frac{a + b}{2}$$

of two numbers a and b, there is another important mean that we can take when the numbers are nonnegative. The *geometric mean* of a and b is defined to be the number

$$\sqrt{ab}.$$

In taking the geometric mean, we have replaced addition with multiplication and division with a square root. Although it is possible to define the geometric mean of more than two numbers, we shall restrict ourselves to two numbers only. The following inequality shows an important relationship between the arithmetic and geometric means:

Inequality of the Arithmetic and Geometric Mean. Given any two nonnegative numbers a and b we have

$$\frac{a + b}{2} \geq \sqrt{ab} \, .$$

It is easy to prove this inequality using rules (1)–(16), and we leave this as an exercise. The inequality can also be pictured geometrically. A glance at Figure IV.3.5 immediately reveals that $(a + b)^2 \geq 4ab$, with equality if and only if $a = b$. This inequality is equivalent to the desired result.

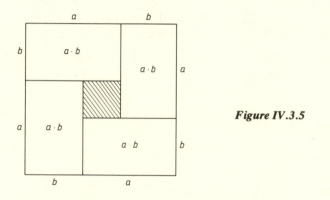

Figure IV.3.5

3.4   The Behavior of Polynomial Functions

In this subsection we shall sharpen our understanding of inequalities by discussing the behavior of polynomial functions. The facts about polynomials derived in this subsection are used daily in every quantitative science.

The simplest kind of polynomial is the function f of the form $f(x) = x^n$ (where n is a nonnegative integer). This function associates to each number x its nth power. When $n = 0$, we understand the function to be the constant function 1. The graphs of the functions f that are obtained this way are shown in Figure IV.3.6 for the first few values of n. Using rules (1)–(16) of subsection 3.2, we can deduce the following properties of these special polynomials:

(a) The function x^n is increasing if we confine ourselves to nonnegative values of x. In other words, as x increases, so does the number x^n. This fact is clear when $n = 1$, and, using rule (6), we may prove the general case by mathematical induction.

(b) If n is even, then $(-x)^n = x^n$. In other words, x^n is symmetric about 0. This is clear

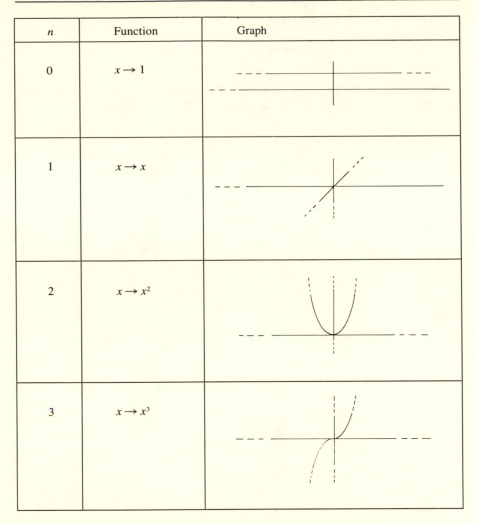

n	Function	Graph
0	$x \to 1$	
1	$x \to x$	
2	$x \to x^2$	
3	$x \to x^3$	

Figure IV.3.6

if $n = 0$ or $n = 2$, and the general case is easy to see using induction. Alternatively we may note that

$$(-x)^n = (-1)(-1) \ldots (-1)x^n,$$

where the factor -1 appears n times. Since n is even, we may pair off the factors -1 and use the fact that each pair of factors $(-1)(-1)$ is just 1.

In general, a function f that satisfies the identity $f(-x) = f(x)$ for every x is said to be an *even function*.

(c) If n is odd, then $(-x)^n = -x^n$. In other words, x^n is antisymmetric about 0. This property follows in the same way that we proved (b). The important fact is that the product of an odd number of factors -1 can be paired off up to the last factor, and is therefore equal to -1.

In general, a function f that satisfies the identity $f(-x) = -f(x)$ for every x is said to be an *odd function*.

(d) If $n \geq 1$, then the expression x^n can be made as large as we like by making x sufficiently large. More precisely, for any number A there is a number x_0 such that

$$x \geq x_0 \Rightarrow x^n \geq A.$$

As a matter of fact, a number x_0 that will do this job is the larger of the two numbers 1 and A, because if $x \geq x_0$, we have

$$x^n \geq x \cdot 1 \cdot 1 \cdot \ldots \cdot 1 \geq x_0 \geq A.$$

(e) If n is even and $n \geq 2$, then x^n can be made as large as we like by making $|x|$ sufficiently large. If n is odd, then x^n can be made as "negatively large" as we like by making x sufficiently "negatively large." More precisely, for any number A there is a number x_0 such that

$$x \leq x_0 \Rightarrow x^n \leq A.$$

This property follows at once from (b), (c), and (d).

In general, a *polynomial* is a function f that can be expressed in the form

$$f(x) = a_0 + a_1 x + a_2 x^2 + \ldots + a_n x^n$$

for every number x where n is a nonnegative integer and the numbers $a_0, \ldots, a_n$ (the *coefficients* of the polynomial) are given. It can be shown that this representation is unique: that is, if we alter one or several of the coefficients $a_0, \ldots, a_n$, the function is no longer the same. Apart from the constant function 0, every polynomial has a last nonzero coefficient, and can therefore be written in the form

$$a_0 + a_1 x + a_2 x^2 + \ldots + a_n x^n,$$

where the coefficient a_n is not zero. If we have done this, then we call $a_n x^n$ the *highest term*, and the number n the *degree* of the polynomial.

Using properties (a)–(e), we can now state the following result:

(f) Suppose that f is the polynomial defined by

$$f(x) = a_0 + a_1 x + a_2 x^2 + \ldots + a_n x^n$$

for each x. Then, provided that $|x|$ is large, $f(x)$ has the same behavior as its highest term $a_n x^n$. We shall give this property a precise statement as we prove it.

We begin by observing that as long as $x \neq 0$ we have

$$f(x) = \left(\frac{a_0}{a_n} \cdot \frac{1}{x^n} + \frac{a_1}{a_n} \cdot \frac{1}{x^{n-1}} + \ldots + \frac{a_{n-1}}{a_n} \cdot \frac{1}{x} + 1 \right)(a_n x^n)$$

Using the rules on inequalities, one may show that the first n terms of the expression

$$\frac{a_0}{a_n} \cdot \frac{1}{x^n} + \frac{a_1}{a_n} \cdot \frac{1}{x^{n-1}} + \ldots + \frac{a_{n-1}}{a_n} \cdot \frac{1}{n} + 1$$

can be made as small as we like by making $|x|$ sufficiently large, and, therefore, if $|x|$ is sufficiently large, we have

$$\left| \frac{a_0}{a_n} \cdot \frac{1}{x^n} + \frac{a_1}{a_n} \cdot \frac{1}{x^{n-1}} + \ldots + \frac{a_{n-1}}{a_n} \cdot \frac{1}{x} \right| < \frac{1}{2} \, .$$

For such values of x we have

$$\frac{1}{2} < \frac{a_0}{a_n} \cdot \frac{1}{x^n} + \frac{a_1}{a_n} \cdot \frac{1}{x^{n-1}} + \ldots + \frac{a_{n-1}}{a_n} \cdot \frac{1}{x} + 1 < \frac{3}{2} \, ,$$

and so we conclude that if $|x|$ is sufficiently large, then $f(x)$ is the product of $a_n x^n$ and a factor that lies between $1/2$ and $3/2$. This is the sense in which we say that $f(x)$ and $a_n x^n$ behave in the same way.

Using property (f), we deduce the following facts:

If $n > 0$, then $|f(x)|$ tends to infinity as $|x| \to \infty$.

If $a_n > 0$, and n is even, then $f(x)$ will be positive whenever $|x|$ is sufficiently large.

If $a_n < 0$, and n is even, then $f(x)$ will be negative whenever $|x|$ is sufficiently large.

If $a_n > 0$, and n is odd, then for $|x|$ sufficiently large we have $f(x)$ positive if x is positive, and negative if x is negative.

If $a_n < 0$, and n is odd, then for $|x|$ sufficiently large we have $f(x)$ negative if x is positive, and positive if x is negative.

An important fact arising from the properties we have just proved is that if the degree of a polynomial is odd, then the polynomial attains both positive and negative values. From an important theorem in analysis known as the *intermediate value theorem*, we now deduce that for every polynomial f of odd degree, there is a number x such that $f(x) = 0$. Such a number x is said to be a *zero* of the polynomial.

§4   *n-Person Games with n ≥ 2*

As we have said, the optimization problems with which we have been concerned up till now can be thought of as *one-person games*. For the rest of this chapter we shall be concerned with *two–person games* and, more generally, with *n-person games*, where n is a natural number and $n \geq 2$. The important thing about a game of this type is that the outcome for each participant does not depend only on his (or her)[1] particular actions. It depends on the behavior of the other players as well. In order to gain a perspective on the many possibilities that may exist in this kind of situation, we shall begin with some concrete examples.

---

[1] From now on, in the interest of brevity, we shall write "he" and "his" instead of "he or she" and "his or her." We do not intend to assign a gender to any of our players.

4.1   Stone-Paper-Scissors

This well-known game is played by two people who have to choose among three strategies: P (for paper), Sc (for scissors), and St (for Stone). The outcomes are determined according to the following rules:

Paper wraps stone.
Stone blunts scissors.
Scissors cuts paper.

The outcomes may be displayed in the form of a *bimatrix* (therefore, the game is a *bimatrix game*). This bimatrix is illustrated in Figure IV.4.1.

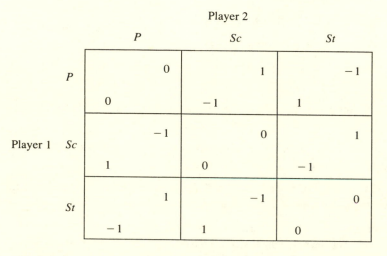

Player 2

		P	Sc	St
		0	1	−1
	P	0	−1	1
		−1	0	1
Player 1	Sc	1	0	−1
		1	−1	0
	St	−1	1	0

Figure IV.4.1

In each of the $3 \cdot 3 = 9$ boxes of the bimatrix we write two numbers, each of which is -1, 0, or 1. The number 1 stands for a gain, 0 stands for a tie, and -1 stands for a loss. The lower left number in each box is the outcome for player 1 who has chosen the row of the bimatrix that contains the given box, and the upper right number in the box is the outcome for player 2 who has chosen the column. We note that the sum of the two numbers in each box is zero, since a gain for one player must be counterbalanced by a loss for the other. A game of this type is said to be a *zero-sum game*. For the complete determination of a zero-sum game, all we need is a simple matrix (rather than a bimatrix). The matrix for the present game is illustrated in Figure IV.4.2. For example, all we have to do is list the outcomes for player 1, and then the outcomes for player 2 may be obtained by changing the signs. For this reason, zero-sum games are also called *matrix games*.

A player who "peeks" can choose an optimal answer against the strategy of his opponent. He can play Sc against P, St against Sc, and P against St, and thus always win. However, if both of the players play honestly, then each has to choose, independently,

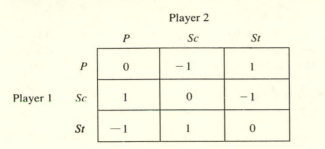

Figure IV.4.2

among the three options in a random fashion. Each of the three options will be chosen with frequency 1/3, and each of the nine possible combinations PP, PSc, . . . , StSt will appear with a frequency of 1/9. Therefore, the average outcome for each player will be

$$\frac{1}{9}(0 - 1 + 1 + 1 + 1 + 0 - 1 - 1 + 0) = 0.$$

Suppose now that player 2 maintains this random $\frac{1}{3} : \frac{1}{3} : \frac{1}{3}$ choice of the options, while player 1 changes to frequencies F_P, F_{Sc}, and F_{St}. Now the average outcome for each player is

$$F_P \cdot \frac{1}{3} \cdot (0) + F_P \cdot \frac{1}{3} \cdot (-1) + F_P \cdot \frac{1}{3} \cdot (1)$$

$$+ F_{Sc} \cdot \frac{1}{3} \cdot (1) + F_{Sc} \cdot \frac{1}{3} \cdot (0) + F_{Sc} \cdot \frac{1}{3} \cdot (-1)$$

$$+ F_{St} \cdot \frac{1}{3} \cdot (-1) + F_{St} \cdot \frac{1}{3} \cdot (1) + F_{St} \cdot \frac{1}{3} \cdot (0)$$

$$= \frac{1}{3} \cdot F_P \cdot ((-1)+1) + \frac{1}{3} \cdot F_{Sc} \cdot ((-1)+1) + \frac{1}{3} \cdot F_{St} \cdot ((-1)+1) = 0.$$

From this we conclude that a player cannot improve his average outcome against the 1/3 mixture of the other player. This 1/3 mixture therefore represents a kind of *equilibrium*.

4.2   The Game of NIM

This well-known game is played by two people according to the following rules:

Let N be a natural number. We distribute N matches in 3 nonempty heaps. The players now move alternatively, each removing some (at least one, and possibly all) of the matches in any *one* of the heaps that happen to be nonempty at that moment. The player removing the last match is the winner.

This game passes through a sequence of states, each of which is fully described by three nonnegative integers n_1, n_2, and n_3 that have starting values N_1, N_2, and N_3. These starting values are positive, and $N_1 + N_2 + N_3 = N$. Each move by one of the players lowers the value of exactly one of the numbers n_i by at least 1. The player arriving at the state 0,0,0 is the winner. The primary question to ask about this game is whether there is a winning strategy. We shall show that the answer is indeed yes. To this end we introduce the notion of a *kernel*.

A set K of states is called a *kernel* if it fulfills the following conditions:

(1) The final state 0,0,0 belongs to K.
(2) No move can lead from a state n_1, n_2, n_3 inside K to a state m_1, m_2, m_3 that is also inside K. In other words, if you are inside K, you *cannot* stay there.
(3) For every state n_1, n_2, n_3 outside K, there is at least one move that leads to a state m_1, m_2, m_3 inside K. In other words, if you are outside K, you *can* get in.

If there is a kernel, then the player encountering a state n_1, n_2, n_3 outside K has a safe, winning strategy G: always move into K. His opponent will find himself in K every time and must therefore be forced to move the game back out of K. Therefore, since 0,0,0 is inside K, the opponent can never leave the game in the state 0,0,0. Since the number of matches in the game decreases every time a move is made, the game must end in a finite number of steps, and therefore the first player must win.

The important question to ask is, therefore,

> Does NIM have a kernel?
> Furthermore, if there is a kernel, describe it!

One can see fairly easily that NIM cannot have *more* than one kernel. To obtain a contradiction, suppose that K and L are both kernels, and that K contains a state n_1, n_2, n_3 which does not belong to L. It is possible to move the game to a new state that does belong to L, but this new state cannot be in K. We can now take this new state to a third state which is in K but not in L. We can continue this process indefinitely without reaching the state 0,0,0, which lies in both kernels. But since every move decreases the number of matches in the game, the matches are all gone after at most N moves, and we therefore have the desired contradiction (see Figure IV.4.3).

We shall now prove the existence of a kernel by giving an explicit description of the only possible kernel. For this purpose we shall expand the numbers n_1, n_2, and n_3 in binary form.

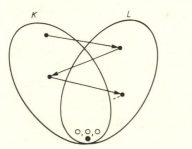

Figure IV.4.3

$$n_1 = a_0 + a_1 \cdot 2 + a_2 \cdot 2^2 + \ldots + a_r \cdot 2^r$$
$$n_2 = b_0 + b_1 \cdot 2 + b_2 \cdot 2^2 + \ldots + b_r \cdot 2^r$$
$$n_3 = c_0 + c_1 \cdot 2 + c_2 \cdot 2^2 + \ldots + c_r \cdot 2^r,$$

where each of the coefficients a_i, b_i, c_i is either 0 or 1. We define K to be the set of all states n_1, n_2, n_3 that have the property that every one of the vertical sums $a_0 + b_0 + c_0$, $\ldots$, $a_r + b_r + c_r$ is an even number. In other words, every one of these sums is either 0 or 2.

Suppose, for example, that the three piles have 1, 3, and 5 matches respectively. Since

$$1 = 1 + 0 \cdot 2 + 0 \cdot 2^2 + \ldots + 0 \cdot 2^r,$$
$$3 = 1 + 1 \cdot 2 + 0 \cdot 2^2 + \ldots + 0 \cdot 2^r,$$
$$5 = 1 + 0 \cdot 2 + 1 \cdot 2^2 + \ldots + 0 \cdot 2^r,$$

the vertical sums are 3, 1, 1, 0, $\ldots$, 0 ; some of these numbers are not even. Therefore, the state 1, 3, 5 is not in K. However, if we replace the largest number, 5, by

$$2 = 0 + 1 \cdot 2 + 0 \cdot 2^2 + \ldots + 0 \cdot 2^r,$$

then we obtain new vertical sums 2, 2, 0, $\ldots$, 0, which are all even, and therefore the state 1, 3, 2 is in K. If we now decrease any *one* of these three numbers, then at least one of the digits 1 in their binary expansions is changed into 0, which changes the corresponding vertical sum from even to odd: we cannot stay in K.

The ideas contained in this numerical example can easily be extended into a general proof that the set K is a kernel. We leave this (not too difficult) task for the reader.

It is possible to write any NIM game in matrix form, but this approach is not very interesting. See Jacobs [1983a] for further information on NIM. The "NIM idea" is of general importance for "numbers and games." See Conway [1976] and Berlekamp-Conway-Guy [1982].

4.3 Prisoners' Dilemma

Prisoners' dilemma is a much investigated bimatrix game; see, for example, Brams [1983]. The players are two prisoners who are accused of a crime that they committed jointly. They are in pretrial custody and cannot exchange information. Each has a choice between A (admitting) or D (denying) the charge. Depending on what option they pick, the court will grant them sentences less than the maximum of 5 years in prison according to the bimatrix shown in Figure IV.4.4.

Player 1 thinks as follows: *No matter what player 2 does, I am better off with A than with D. Therefore, I choose A.* Player 2 reasons in the same way and also chooses A.

Now they both have second thoughts: *My pal has the same thoughts that I have. Like me, he will come to the conclusion that we are both better off if we both choose D. Anticipating this, I will change my choice to D.*

Now player 1 thinks again: *Player 2 has changed his choice to D. Therefore, I will be better off if I choose A . . .*

If we look at the two players as superpowers, and interpret A as dove tactics and D as hawk tactics, then we arrive at a very rough picture of the present political situation. In

Player 2

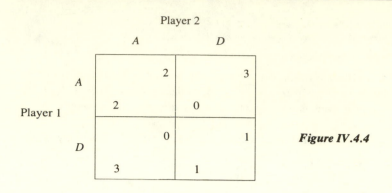

Figure IV.4.4

fact, game-theoretical considerations have played a role in the background of superpower diplomacy for decades. See Brams [1985].

4.4   Some More Bimatrix Games

In the *revelation game*, player 1 is a superior being who has the choice between R (revealing) and C (concealing) himself, and player 2 is a man with the options B (believing) or N (not believing) in the existence of the superior being. The feelings of the two players are summarized in the bimatrix in Figure IV.4.5. These feelings may be expressed verbally as follows:

(1) To the superior being, a man who believes in spite of nonrevelation represents the highest value (4). Belief as a consequence of revelation is preferred (3) to nonbelief in the absence of revelation (2). Nonbelief in spite of revelation is a disgrace (1).

(2) For the human player, belief as a consequence of revelation is most satisfactory (4). The stubborn "no revelation, no belief" is preferred (3) to blind belief (2). Nonbelief in spite of revelation is stupid (1).

Situation C, N (no revelation, no belief) is an equilibrium in the following sense: If

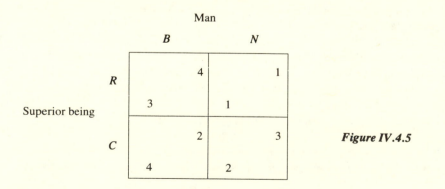

Figure IV.4.5

the superior being refuses to reveal himself, then passage to belief is unsatisfactory to the man $(2 \leftarrow 3)$. Conversely, if the man insists on disbelief, then the superior will lose $(2 \rightarrow 1)$ only if he offers revelation cheaply.

Newcomb's problem deals with the following situation: Player 1 (again a superior being) puts a thousand dollars into box number I and has the choice (E) to leave box number II empty, or (M) to put a million dollars into it. Player 2 is a man who has two options: empty both boxes (B), or empty box II only (II). Figure IV.4.6 gives the possible outcomes for the man. (It is, of course, more profitable for him to empty both boxes, no matter what the superior being does.)

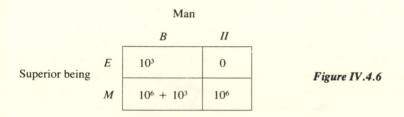

Figure IV.4.6

So far, everything is clear. Newcomb's problem arises as soon as the superior being tries to *predict* what player 2 will do, and makes his decision dependent upon this prediction in the following manner:

prediction B: E (greed is punished)
prediction II: M (modesty is rewarded)

Now the total situation can be put into the form of the bimatrix shown in Figure IV.4.7. The lower left number in each box represents the satisfaction (1) or the disappointment (0) of the superior being, depending upon the fulfillment or nonfulfillment of the prediction.

This bimatrix game has been discussed in many publications. In order to gain an impression of the problems involved, imagine the following scenario:

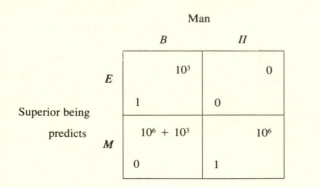

Figure IV.4.7

Player 2 (the man) reasons as follows: "If the superior being is omniscient, then it knows what I will do. Therefore, we will always end up in the diagonal of the bimatrix, and therefore I will be better off with II. However, if the superior being is not omniscient but believes that I consider it to be omniscient, then he will imagine my previous thought and put a million dollars in box II. But then I can play a trick on him by opening both boxes and pocketing an extra thousand dollars."

A broad discussion of this omniscience problem can be found in Brams [1983].

4.5   Game Theory and Biology

During the past two decades, game theory has invaded biology. More precisely, it has invaded the field of ethology and evolutionary theory. There now exists a "Darwinism on the gene level," which can be used to explain certain phenomena (like unselfishness) that used to be viewed as hints of the existence of a divine creator. The books by Smith [1982] and Hofbauer-Sigmund [1988] are standard references, and the nonmathematician will find Dawkins [1976] and Wickler-Seibt [1977] especially useful. In this subsection we shall discuss just one example, the *hawk-dove game*.

The hawk-dove game, represented by the bimatrix in Figure IV.4.8, involves two individuals of the same species who are engaged in a fight. (The terms "hawk" and "dove" are, of course, borrowed from political jargon.) Each has a choice between two strategies:

H (hawk): fight until you win or are wounded.
D (dove): test your opponent and flee if he remains obstinate.

The prize for winning the fight is food with a value V (to be measured on some biological scale). If hawk meets dove, then dove gets nothing and hawk gets everything. If dove

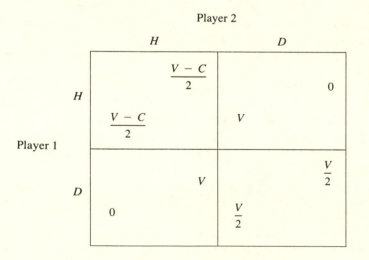

Figure IV.4.8

meets dove, they share fifty-fifty. If hawk meets hawk, they also share fifty-fifty, but they each lose $C/2$ (C is the cost) in loss of energy and wounds from the fight.

The value of the two strategies depends upon the relationship between V and C. If $V \geq C$, then H is better in all cases. If $V < C$, then being an H is better against a D, but being a D is better if you meet another D.

We shall now look at this game from the standpoint of evolutionary theory. Imagine a large population of a species in which a certain fraction p is genetically programmed to be H, and a fraction $1 - p$ is programmed to be D. This fraction p may change over time because of mutation, proliferation, and selection. The values V and C are measured in degrees of survival fitness. We shall now look at the cases $V > C$ and $V < D$.

Case I: $V > C$. Any individual who is born as an H has an advantage. The H-genes conquer the whole gene pool, and after a while, a pure H-population results.

Case II: $V < C$. In a population with a fraction p of H's and a fraction $1 - p$ of D's, an individual who is an H will meet with an H with frequency p, and will meet a D with frequency $1 - p$. Each time an H meets another H it receives $(V - C)/2$. We note that this number is negative, which means that the individual suffers a loss in each of these encounters. Each time it encounters a D, it receives V. Therefore, on average it receives

$$p \left(\frac{V - C}{2} \right) + (1 - p)V.$$

For an individual who is a D, the corresponding average is

$$p \cdot 0 + (1 - p)\frac{V}{2}$$

To determine which of the two is better off, we calculate the difference

$$p \left(\frac{V-C}{2} \right) + (1-p)V - (1-p)\frac{V}{2} = p\frac{V}{2} - p\frac{C}{2} + V - pV - \frac{V}{2} + p\frac{V}{2} = \frac{1}{2}(V - pC).$$

The latter expression is zero (and the system is in equilibrium) when $p = V/C$. When $p > V/C$, the difference is negative, the D's are better off than the H's, and p (the proportion of H's) decreases. Similarly, when $p < V/C$, the H's are better off and p increases. We conclude that the proportion p of H's will eventually approach the stable value of V/C.

Using this type of argument, Ronald Alymer Fisher (1890–1962) had already shown by 1930 that a proportion $\frac{1}{2} : \frac{1}{2}$ among the sexes is evolutionarily stable for humans, even though a relatively small number of males would be sufficient to mate with all females. The concept of an ''evolutionarily stable mixed strategy'' is fundamental in evolutionary game theory (see Subsection 6).

4.6   Aggregation of Preferences and Arrow's Dictatorship Theorem

A person confronted with the choice ''coffee (C) or tea (T)'' would immediately set up a *preference order* for the two choices. The two possibilities are coffee preferred to tea: $\begin{array}{c} C \\ T \end{array}$

tea preferred to coffee: $\begin{array}{c} T \\ C \end{array}$

Now suppose that two people are presented with the choice "coffee or tea," and they have to reach a consensus on which beverage to order. To reach this consensus, they need to derive or *aggregate* a common preference order from their individual preference orders. In general, problems of this type arise with a number $n \geq 2$ of individuals who are each confronted with a number $m \geq 2$ of *options*. We shall see that when $m = 2$, well-established aggregation procedures are available: we allow the majority to rule. However, if $m \geq 3$, then we will find ourselves facing some unwelcome mathematical theorems, such as *Arrow's dictatorship theorem*.

Returning now to our coffee-tea problem, the unanimous patterns

$$\begin{array}{cc} \underline{1} & \underline{2} \\ C & C \\ T & T \end{array} \quad \text{and} \quad \begin{array}{cc} \underline{1} & \underline{2} \\ T & T \\ C & C \end{array}$$

are easy to handle. The pattern on the left results in C, and the pattern on the right results in T. In the controversial situations

$$\begin{array}{cc} \underline{1} & \underline{2} \\ C & T \\ T & C \end{array} \quad \text{and} \quad \begin{array}{cc} \underline{1} & \underline{2} \\ T & C \\ C & T \end{array} \,,$$

we could solve the problem fairly by calling for C in both cases. (Then every person wins once: person 1 wins in the left pattern and person 2 wins in the right pattern.) Calling for T in both cases would be equally good. On the other hand, C in the left and T in the right would allow person 1 to win in both cases and thus become a "dictator."

Whenever we have to aggregate patterns of n individual preference orders for just *two* options, we have a well-established method: majority vote, plus decision by a chairman in case of a tie. If $n \geq 3$, then the chairman can be outvoted, i.e., he is not a dictator. If $n = 2$, then the chairman is a dictator, but in a case like this we can call on "ethics." For example, we could decide in advance which preference order should be adopted should a controversy arise.

Things become difficult only when there are three or more options. We shall now add whiskey (W) as a third option. For two people presented with these three options, we have patterns like

$$\begin{array}{cc} \underline{1} & \underline{2} \\ C & T \\ T & W \\ W & C \end{array}$$

According to this pattern, both people agree about T and W in that they prefer T to W. But they disagree strongly about C.

This is the kind of difficulty that arises when a faculty sits around a table to elect a dean, and every person votes for his left neighbor. The book by Black [1958] reports on the mathematical literature on such themes up to 1949. Other books on the topic include those by Straffin [1980] and Kenneth Arrow [1951]. Arrow published the following result, now often called "Arrow's paradox," in 1951. He was awarded the Nobel Prize for Economics in 1972.

> Every aggregation method for $n \geq 2$ individuals confronted with $m \geq 3$ options, which satisfies certain plausible conditions, is automatically dictatorial.

When we say that a method is "dictatorial," we mean that among the n people there is one such that the method simply results in the preference order of this particular person. This person is the dictator for the given method. Clearly, there are n such methods, corresponding to the n possibilities of choosing the dictator. We shall now state the two "plausible conditions" that Arrow's paradox requires to guarantee that an aggregation method is dictatorial:

> (1) Unanimity rule: For any two of the m options ($m \geq 3$)—call them C and T—in which all n people agree on "C above T," the method must result in a preference order that sets C above T.
>
> (2) Independence rule: For any two of the m options ($m \geq 3$)—call them C and T—if the n people change their preferences without changing their order of preference for the options C and T, then the order of the options C and T in the aggregate is also unchanged.
>
> In other words, if no person changes his mind on the order of C and T, then collectively they do not change the order of C and T.

We can now restate Arrow's paradox more precisely.

Arrow's Dictatorship Theorem. There are precisely n aggregation methods fulfilling the unanimity rule and the independence rule, namely

> method 1: individual 2 dictates
>
> . . .
>
> method n: individual n dictates.

The proof of this theorem is not too difficult, but we do not have the space to present it here. See, for example, Black [1958], Straffin [1980], Jacobs [1983a], and the literature cited in these works. Among the items in this literature, we particularly mention Peleg [1978], [1984]. Bezalel Peleg (Jerusalem) was able to smooth out the awkward situation generated by Arrow's paradox to some extent. We end this subsection with a short review of the historical development of Arrow's paradox.

The way Arrow formulated his paradox in 1951 is a little different from the way it has been presented here. He stated that the three conditions

> (1) Unanimity rule
> (2) Independence rule
> (3) No dictator

are logically incompatible. This formulation is, of course, logically equivalent to the one given here. Since Arrow's paradox is a rigorous mathematical theorem, there is only one way to avoid its conclusion: change the hypotheses.

A first attempt in this direction involved weakening the requirements that had been imposed on the aggregation mapping. It no longer demanded the distillation of a whole preference order out of the pattern of individual preference orders, but only of one single top preference. This approach opens up new possibilities for the individual player. While it had been difficult for the individual to evaluate the aggregated preference order in comparison with his own preference order, he could now find the aggregated top preference somewhere on his own list, and he could measure his satisfaction or dissatisfaction with the aggregated outcome accordingly. The individual might then try to "bluff," that is, use a *false* preference order as his input into the aggregation machinery, in order to influence the outcome favorably with respect to his *true* preference order. Unfortunately, this theoretical attempt ended with one more disappointing theorem: Bluff-proof aggregation methods of the modified kind just described are always dictatorial. This result is known as the Gibbard-Satterthwaite theorem (Gibbard [1973], Satterthwaite [1975]). But Peleg [1978] was able to present a dictator-free aggregation method, which may be described roughly as follows:

(1) Every player has his turn to eliminate the option least desired by him.
(2) The players obey some final decision of a "soft" umpire.

See also Peleg [1984].

§5   *Equilibrium*

In the preceding section we saw in a series of two-person, or even *n*-person, games what effect the collective behavior of all the players may have on a single player, and how the single player might react to this effect. Whenever it is possible to represent the performance of a single player by some place on an ordered ladder, the player will try to move up that ladder. A situation in which no player can do this is called an *equilibrium*. Putting this a little more precisely, but still heuristically, we can handle equilibrium as follows:

We say that an *n*-person game is in *equilibrium* if the strategies of the single players are such that no single player can improve his situation by changing his strategy as long as all the other players keep their strategies fixed.

We encountered several examples of equilibria in Section 4:

The paper-stone-scissors game is in equilibrium if the two players play with frequencies 1/3, 1/3, 1/3.

In NIM we encountered a definite winner-loser situation.

The prisoners' dilemma game is in equilibrium if both players deny the charge.

The revelation game is in equilibrium if the superior being does not reveal himself and the man does not believe.

The hawk-dove game with $V < C$ is in equilibrium when the proportion of hawks is V/C.

The only interesting variations of Newcomb's paradox and Arrow's paradox do not provide situations in which the game is in equilibrium. However, such situations arise in connection with the Gibbard-Satterthwaite version of Arrow's paradox.

In the present section we shall consider n-person games for which it is perfectly clear what equilibrium means. The only question that remains is what hypotheses will guarantee that an equilibrium exists. A precise mathematical answer to a question such as this is called an *equilibrium theorem*. In this section we shall present the most important equilibrium theorems. The first of these deals with what we call *tree games* and is of a combinatorial nature because it deals with only a finite number of possibilities. Our second equilibrium theorem deals with what we call *noncooperative n*-person games and makes use of a concept known as "mixed strategy." The proof of the theorem is topological in nature; it is based on the fixed-point theorem of Brouwer and Kakutani (see Chapter V, Section 5). The same is true of the equilibrium theorems of mathematical economics, which are presented succinctly in Subsection 3. Nonmathematicians will find a good overall introduction to game theory in Davis [1970]. To mathematical beginners, the book by Franklin [1980] is especially recommended.

5.1   The Equilibrium Theorem for Tree Games

We shall begin with a simple tree game that is played everywhere. Player 1 is a citizen, player 2 is the police, and player 3 is justice.

Step 1: Player 1, being observed by the police, chooses one of several possible types of behavior.

Step 2: The police, realizing the choice made by player 1, decides between the two possibilities "get active" and "don't get active." In the latter case, the game ends with outcomes of 0 for all players.

Step 3: The police, having decided to "get active," now chooses one of several ways of doing so.

Step 4: Player 1, confronted with the action of the police, decides how to react to it. In some cases the game now ends with an outcome of -2 for the police (cost of efforts) and a negative outcome for player 1 (a fine), while justice, who is not involved, receives 0. In all other cases the game proceeds to step 5.

Step 5: Justice becomes involved and decides on one of several possible courses of action.

. . .

The game continues in this way, possibly through several stages of appeal. In each case the game ends at some point, and the outcomes for the players can then be realized. We shall assume that every outcome can be specified in the form of a number. The different possibilities in the game can be illustrated in the form of a tree diagram as shown in Figure IV.5.1.

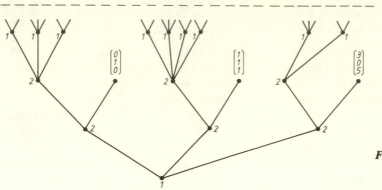

Figure IV.5.1

The game moves upward along a certain path, starting from the root of the tree. At every branching point, the player whose turn it is to play—his number is attached to that node—chooses the next branch of the path. The game ends as soon as the path reaches a tip of the tree. At this tip we read the three outcomes for the three players, arranged vertically.

Among the various kinds of people who might play this game, there are those, the "cranks," who have long since arranged what they will do at every node of the tree at which they will play. Actually, these "cranks" are the ideal players in the sense of game theory, because they have a *strategy* that tells them what to do in every imaginable situation. Adopting this ideal, we shall assume that every one of the three players has a pool of strategies from which he chooses one. If all of the players have chosen their respective strategies, then the game is completely determined. It runs from the root of the tree to one of its tips along a certain path, and it results in a single numerical outcome (a "payoff") for each player. In much the same way, we can describe a game with n players by means of a tree that has been suitably labeled with the number of players and the outcomes. In this event, the n strategies chosen by the n players will result in n payoffs, one to each player.

Assume that the strategies of the three players are made known as soon as all of the choices have been made. Each player now knows the choices of the $n - 1$ others in the game, and may scan all these possible strategies with the view of improving his own payoff. Naturally, this makes sense only if all the other players are going to stick to their strategies, and he alone is allowed to make changes. If he cannot improve his payoff, he has no reason to change his own strategy. In the event that all n players come to this conclusion, then we shall say that the given choice of n strategies represents an *equilibrium*.

An interesting question arises at this point: *can we be sure that an equilibrium exists?* The answer to this question is yes, and is known as the *equilibrium theorem for tree games*.

Equilibrium Theorem for Tree Games. Every tree game has at least one equilibrium (Kuhn [1950]).

Proof. We prove this theorem by induction on the height N of the tree. When $N = 0$, the tree consists only of its root, and no player can do anything. In other words, every player has exactly one strategy, "do nothing," and the game is in equilibrium. When $N = 1$, the tree has the form shown in Figure IV.5.2. Player k is the only player who has any real choice. He has to choose one of the branches. The other players have only one strategy each: do nothing. If player k chooses a path giving him a maximal payoff, then the game is in equilibrium.

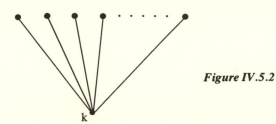

Figure IV.5.2

Assume now that the theorem is true for all trees of height not exceeding $N - 1$. Consider a game whose tree has height N. We illustrate this tree in Figure IV.5.3. Player k begins and has a choice among r branches. Once he has decided, the remainder of the game evolves in the subtree resulting from his decision. Each of these subtrees has a height not exceeding $N - 1$, so each subtree must have at least one equilibrium. There may be many possible equilibria in a given subtree, but in each subtree we shall choose just one. We shall call the equilibrium that we have chosen in tree j "equilibrium number j." Now player k looks at his payoff in equilibrium number j for each j, and he selects subtree j_0 in such a way as to make his payoff as large as possible. We have thus described the overall strategy of player k: decide for subtree j_0 at the root of the tree, and for each j, play according to his strategy as described in equilibrium number j. Actually, he need worry only about his strategy in subtree $j = j_0$, but for the sake of the general definition

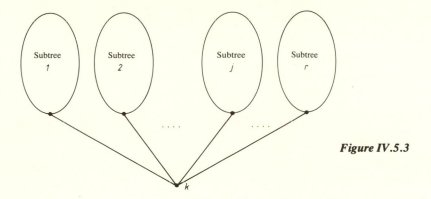

Subtree 1   Subtree 2   Subtree j Subtree r

Figure IV.5.3

we describe his strategy in the other subtrees as well. Every other player also combines his strategy in each of the subtrees to make an overall strategy on the main tree.

We have now obtained a choice of overall strategies that constitutes an equilibrium for the original tree game. No player can improve his lot if the others stick to their choices.

We shall now give an example to demonstrate the possibility that a game may have several equilibria. See Figure IV.5.4. In every subtree, only player 2 has to make a choice, and his payoff in the subtree does not depend on how he chooses. If he chooses "left" in both subtrees, then player 1 will decide for subtree number 1, and the final equilibrium payoffs are $\begin{pmatrix} 3 \\ 0 \end{pmatrix}$. If player 2 decides on "right" in every subtree, then player

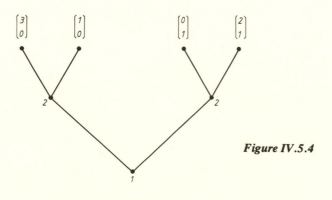

Figure IV.5.4

1 will decide for subtree number 2, and the final equilibrium payoffs will be $\begin{pmatrix} 2 \\ 1 \end{pmatrix}$.

Our assumption that all players make their strategies public after all the choices have been made is crucial to the validity of the equilibrium theorem. If we remove this assumption, then we must make fundamental changes in the notion of strategy. In this new situation, each player has to make his decisions possibly without the knowledge of precisely where in the tree (if anywhere) the decisions will be implemented.

We now give an example of this kind of game. In this example, the IRS and the taxpayer (Selten [1982]), we also take the opportunity to display the phenomenon of random moves in a game. The game has three players:

> player 0, the random player,
> player 1, the IRS,
> player 2, the taxpayer.

Because none of the players know what the others are about to do at any stage of the game, we have to partition the nodes of the tree of this game into what are called "information districts" $\begin{bmatrix} - & - & - \\ - & - & - \end{bmatrix}$.

For each node in a given information district, we need to offer the same options to the player whose turn it is to move. Using these options, the player will make his decision in the same way at each node in the district. Each player therefore sets his strategy simply by making his decisions for each information district that applies to him. Now consider the diagram shown in Figure IV.5.5.

In this figure, the payoffs are shown with the upper number representing the payoff of player 1 and the lower number representing the payoff of player 2. The occurrence of random moves causes random arrivals at the tips of the tree, and therefore causes random

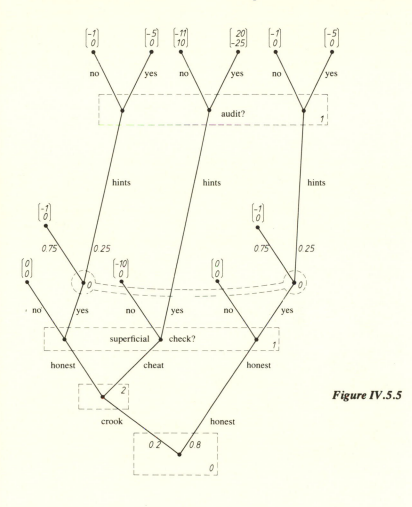

Figure IV.5.5

payoffs at those tips. If, for example, the IRS (being made up of stubborn bureaucrats) always performs audits, and if the taxpayer is a crook who cheats on his returns, than all nonrandom decisions go to the right, and the game ends up with probability

$$0.2 \text{ at } \begin{pmatrix} 20 \\ -25 \end{pmatrix} \text{ (crook punished)}, (0.8)(0.75) = 0.6 \text{ at } \begin{pmatrix} -1 \\ 0 \end{pmatrix} \text{ (inconclusive}$$

$$\text{tax audit at the IRS's cost), and } (0.8)(0.25) = 0.2 \text{ at } \begin{pmatrix} -5 \\ 0 \end{pmatrix} \text{ (thorough but}$$

inconclusive tax audit at the IRS's cost).

Combining these, we obtain an expected payoff for the IRS of

$$(0.2)(20) + (0.6)(-1) + (0.2)(-5) = 2.4,$$

and for the taxpayer,

$$(0.2)(-25) + (0.6)((0) + 0.2)(0) = -5.$$

We obtained these payoffs using the standard product formula for the probabilities of independent random variables. The results we have obtained allow us to make the following conclusion:

> *When stubborn bureaucrats vie with determined crooks, the bureaucrats win on the average.*

This situation can change, of course, if other strategies are used, or if we change the payoffs. We may even imagine the IRS making random decisions on what to do, leading to the concept of *mixed strategy*, a central feature of the next subsection.

For many real-life situations, the only realistic models are games in which the players have incomplete information. We all know from daily experience how many decisions we have to make without advance knowledge of the conditions under which the decisions will be implemented. We usually do not even know the payoffs. Therefore, the clean-cut situation we assumed for the preceding equilibrium theorem is the exception, not the rule.

5.2   The Equilibrium Theorem for Noncooperative Games

In section 4 we became acquainted with several bimatrix games. For example, in the "prisoners' dilemma" we considered the bimatrix shown in Figure IV.5.6. Using the bimatrix, we deduced the payoffs of the two players depending on their decisions either to A (admit) or D (deny) the charge against them. The bimatrix shown in Figure IV.5.6. does not permit any equilibrium. In the bimatrix shown in Figure IV.5.7, the

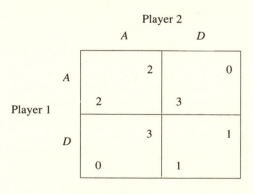

Figure IV.5.6   **Figure IV.5.7**

arrows tell us how the game would oscillate if the two players alternatively attempted to improve their situations, each basing his change on the decision that the other has just made. The bimatrix shown in Figure IV.5.8 for "paper-stone-scissors" leads to a periodic diagram as well. However, in the paper-stone-scissors game, we found an equilibrium-generating device. We let both players act at random and independently with probabilities 1/3, 1/3, 1/3 for the three possible actions. In fact, we showed in Section 4 that if one player acts in this way, then the other has no reason to deviate from the "mixed strategy" 1/3, 1/3, 1/3.

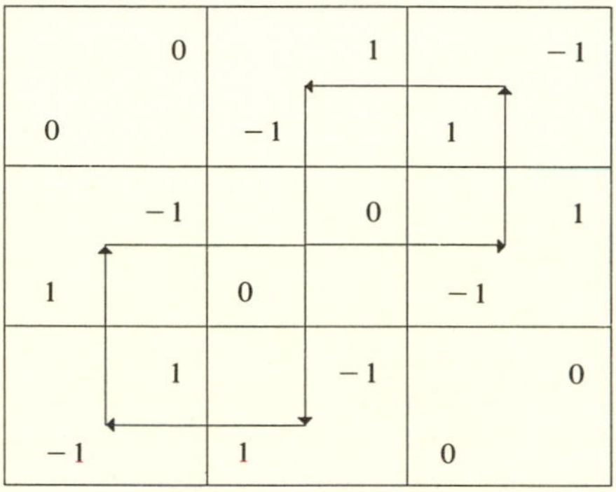

Figure IV.5.8

Our next equilibrium theorem is simply a generalization of this very special example to arbitrary n-person games in which every player has a finite number of possible strategies. For simplicity of notation, we shall show this theorem in the case $n = 3$. The extension to larger values of n will then be clear. Assuming that $n = 3$, we have the following situation:

Player 1 has a choice between his strategies numbered $1, 2, \ldots, r$.
Player 2 has a choice between his strategies numbered $1, 2, \ldots, s$.
Player 3 has a choice between his strategies numbered $1, 2, \ldots, t$.

Note that in this description, strategy number 1 for player 1 may be completely different from strategy 1 for one of the other players.

A *mixed strategy* for player 1 simply means a distribution of nonnegative weights u_1, $\ldots, u_n$ such that $u_1 + \ldots + u_n = 1$. By this we mean that player 1 will use each strategy i with probability u_i. We shall refer to a mixed strategy of this form more briefly as the mixed strategy u for player 1. In the same way, we may speak of a mixed strategy v for player 2 consisting of weights $v_1, \ldots, v_s$, and we may speak of a mixed strategy w for player 3. We note: unless $r = 1$, $s = 1$, and $t = 1$, since the weights u_i, v_j, and w_k may vary continuously, there is a wide variety of possible mixed strategies. In contrast to these mixed strategies, we call the original strategies $1, 2, \ldots$ of a given player the *pure* strategies of that player.

Once the three players have chosen their three mixed strategies u, v, w, we imagine the following procedure: The game is played with pure strategies over and over again in a long sequence of repetitions. In this sequence, player 1 plays his pure strategy number 1 with probability u_1, his pure strategy number 2 with probability u_2, and, in general, his pure strategy i with probability u_i. Players 2 and 3 act in a similar manner, using probabilities from v and w. Each player operates randomly and independently of the other two. We may therefore use the well-known product formula for the probabilities of independent random variables to conclude that in any single game, the probability that player 1 will use his strategy number i while player 2 uses his strategy number j and player 3 uses his strategy number k is $u_i v_j w_k$. If the payoffs received by the three players in this event are A_{ijk}, B_{ijk}, and C_{ijk} , then the average payoffs received by the players are

$$\sum_{i,j,k} A_{ijk} u_i v_j w_k,$$

$$\sum_{i,j,k} B_{ijk} u_i v_j w_k, \text{ and}$$

$$\sum_{i,j,k} C_{ijk} u_i v_j w_k,$$

respectively. Each of the sums

$$\sum_{i,j,k}$$

runs over all possible triplets i,j,k, and therefore contains rst terms. We shall write these three averages more briefly as $A(u,v,w)$, $B(u,v,w)$, $C(u,v,w)$.

An *equilibrium in mixed strategies* is a particular choice $\bar{u},\bar{v},\bar{w}$ of mixed strategies such that no player can improve his payoff by changing his mixed strategy as long as the other two make no changes. For example, if $\bar{u},\bar{v},\bar{w}$ is an equilibrium, and player 1 changes from $\bar{u}$ to a mixed strategy u, then

$$A(u,\bar{v},\bar{w}) \leq A(\bar{u},\bar{v},\bar{w}).$$

The natural question that arises now is whether an equilibrium must always exist. The answer is *yes*, as we see in the following theorem of Nash.

Equilibrium Theorem of Nash [1950], [1951]. In any n-person game of the kind described in this subsection, there is at least one equilibrium in mixed strategies.

Since the players act independently of one another in games of this type, and do not cooperate with each other, these games are called *noncooperative*. The theorem that we have just stated is usually called Nash's equilibrium theorem for noncooperative n-person games.

The proof of Nash's theorem requires some very specialized techniques, and we cannot present it in its complete form here. Roughly speaking, it is performed by setting up a *dynamical system X, T* that fulfills the hypotheses of a result known as *Brouwer's fixed-point theorem*, which we shall discuss in Section 5 of Chapter V. Brouwer's theorem tells us that as long as X is convex and compact and T is continuous, then X must contain a point $\bar{x}$ such that $T\bar{x} = \bar{x}$. Such a point x is said to be a *fixed point* of the system. In this particular case, X is chosen to be the set of all possible combinations $u, \ldots, w$ of mixed strategies of the n players. This space X can be shown to be both convex and compact. Now T is chosen to represent the tendency of every player to improve his situation against the known strategies of the other players. If T is constructed in the right way, it becomes a continuous mapping, so that Brouwer's theorem may be used to find a fixed point $\bar{x}$. $T\bar{x} = \bar{x}$ means that no player, acting alone against the strategies of the other players, can really improve his payoff. That is, $\bar{x} = (\bar{u}, \ldots, \bar{w})$ constitutes an equilibrium.

We end the subsection with some further remarks.

(a) Pure Strategies and Mixed Strategies. If a player has pure strategies numbered 1 to r, then each of these pure strategies can also be identified as a mixed strategy. For example, pure strategy i is the mixed strategy in which the weights $u_1, \ldots, u_r$ have been chosen in such a way that

$$u_i = 1$$
$$u_j = 0 \text{ whenever } j \neq i.$$

Since this choice of weights causes the player to adopt strategy i all the time, the corresponding mixed strategy is simply the pure strategy i. A little less precisely we might then say that *pure strategies are special cases of mixed strategies*, and that the game with mixed strategies is the *mixed extension* of the game with pure strategies. Nash's theorem can now be restated as follows:

> *There may be no equilibrium in pure strategies, but in the mixed extension there is always an equilibrium.*

(b) The Employment of Pure Strategies in an Equilibrium. As soon as Players 2, . . . ,n have chosen their (mixed) strategies, player 1 can scan his pure strategies for the payoffs that they yield against the strategies that the other players have already chosen. He optimizes his total payoff by choosing one of his mixed strategies that yields a maximal payoff. Any other such mixed strategy would yield the same payoff. When the game is in equilibrium, each player has chosen his best possible strategies in such a fashion that he gives nonzero weights only to those of his pure strategies that yield optimal payoffs against the strategies of the other players. Furthermore, the latter pure strategies all have the same payoff. This result is sometimes called the Bishop-Cannings theorem.

(c) The Case of a 2-Person Zero-Sum Game: Minimax. The historical origin of the theory that culminates in Nash's equilibrium theorem was the special case of a 2-person zero-sum game. If player 1 has possible strategies $1, 2, \ldots, r$ and player 2 has possible strategies $1, 2, \ldots, s$, then, after a decision by player 1 to use strategy i and a decision by player 2 to use strategy j, the resulting payoffs may be expressed in the following form:

$$A_{ij} \text{ for player 1,}$$
$$B_{ij} \text{ for player 2.}$$

The condition that the game is a zero-sum game means that $A_{ij} + B_{ij} = 0$ for all i and j. In other words, $B_{ij} = -A_{ij}$. What this says is that after each play, whatever one of the player wins is paid for by the other player's loss. In economics this kind of situation can arise when two groups of people (the players) compete for a commodity that is available only in limited quantities. If player 1 uses a mixed strategy u and player 2 uses a mixed strategy v, the we obtain the ''mixed'' payoffs

$$A(u,v) = \sum_{i,j} A_{ij} u_i v_j$$

for player 1, and $-A(u,v)$ for player 2.

Two given choices $\bar{u}$ and $\bar{v}$ of u and v form an *equilibrium* if

$$A(u,\bar{v}) \leq A(\bar{u},\bar{v}) \text{ for every mixed strategy } u \text{ of player 1}$$

and

$$-A(\bar{u},v) \leq -A(\bar{u},\bar{v}) \text{ for every mixed strategy } v \text{ of player 2.}$$

The first of these statements can be written in the form

$$A(\bar{u},\bar{v}) = \max_{u} A(u,\bar{v}),$$

and the second one can be written as

$$-A(\bar{u},\bar{v}) = \max_{v} (-A(\bar{u},v)).$$

Pulling the minus sign and inverting the inequalities, we can rewrite the latter statement as

$$A(\bar{u},\bar{v}) = \min_{v} (A(\bar{u},v)).$$

For every choice of u' and v', we have the inequality

$$\min_{v} A(u',v) \leq A(u'v') \leq \max_{u} A(u,v'),$$

because passage to a minimum brings us down and passage to a maximum brings us up. Therefore, however strategies u' and v' are chosen, we have

$$\min_{v} A(u',v) \leq \max_{u} A(u,v'),$$

and so we deduce that

$$\max_{u} \min_{v} A(u,v) \leq \min_{v} \max_{u} A(u,v).$$

This result is known as the *minimax inequality*. It is easy to prove and is not a deep mathematical statement. It is also easy to remember because it says that starting with max cannot be worse than starting with min.

Now choose $\bar{u}$ and $\bar{v}$, as before, such that $\bar{u},\bar{v}$ constitute an equilibrium for the game. Then we have

$$\max_{u} \min_{v} A(u,v) \geq \min_{v} A(\bar{u},v) = A(\bar{u},\bar{v})$$

$$= \max_{u} A(u,\bar{v}) \geq \min_{v} \max_{u} A(u,v).$$

We have therefore shown that the reverse of the minimax inequality also holds. Combining these two inequalities we obtain

$$\max_{u} \min_{v} A(u,v) = \min_{v} \max_{u} A(u,v).$$

Furthermore, each side of the latter equation is equal to $A(\bar{u},\bar{v})$ for every equilibrium $\bar{u},\bar{v}$. This equation is the well-known *minimax theorem* of John von Neumann (1903–1957). Von Neumann proved the theorem in 1928, and thereby originated the modern theory of mathematical economics (Neumann [1928]). Several different proofs have been found of this special case of Nash's equilibrium theorem, and some of these proofs make use of mathematical tools that are much more elementary than Brouwer's fixed-point theorem. Until now, no one has found a proof of the general form of Nash's theorem that avoids this heavy tool. We mention that the minimax theorem can be considered to be a part of the linear optimization theory that we discussed in Sections 1 and 2. Using the techniques of those sections, one may find the equilibria in a zero-sum game much more easily than they are found in the proof of Nash's theorem. For example, one may use the simplex algorithm. It is also a remarkable fact that all equilibrium payoffs in a zero-sum game are equal, even though they may vary in the general bimatrix case. An example of this variation was given in Subsection 1.

(d) Evolutionary Stable Strategies (ESS). A typical biological game consists of competition between the individuals of a given species who have a list of possible strategies $1,2, \ldots , r$ available to them, and who encounter each other in two-person situations. Mixed strategies $u = u_1, u_2, \ldots , u_r$ are employed by proportions of a large population. By this we mean that for each $i = 1,2, \ldots , r$, a fraction u_i of the individuals employs strategy i. As these games proceed, payoffs of the special form $A(u,u)$ are of special interest. If a mutant strain with pure strategy i appears in the population, it receives a payoff (in the form of a degree of biological fitness) equal to $A(i,u)$. In the event that $A(i,u) > A(u,u)$, this mutant strain has an overproportional fitness, and thus the proportion in the population of such individuals will increase. Thus in the case of an equilibrium, we have $A(i,u) \leq A(u,u)$ for all pure strategies i. However, this necessary condition is not sufficient to guarantee that the population is in equilibrium. It is still possible to have a pure strategy i with $A(i,u) = A(u,u)$ such that an increase in the proportion of u_i individuals in the population would not decrease the total fitness of the population. Thus, fluctuations might arise. In order to exclude this possibility, we impose the additional condition

$$A(i,u) = A(u,u) \Rightarrow A(i,i) < A(u,u).$$

This condition means that the individuals of this strain would come off worse against individuals of their own type than against the population as a whole. If this condition holds, then an increase in the proportion u_i of these individuals would cause them to meet their own kind more frequently, and the fitness $A(u,u)$ of the total population would decrease. Therefore, a return to the original u would be more successful and would take place. These biological models motivate the following definition:

Definition. Suppose that a given bimatrix game has the same strategy sets for both players and a symmetric payoff. In other words, $A_{ij} = B_{ij}$ for all i and j, where A is the payoff for player 1 and B is the payoff for player 2. A mixed strategy u in the game is said to be *an evolutionary stable strategy* (ESS) if u is an equilibrium, and if for every pure strategy i we have

$$A(i,u) = A(u,u) \Rightarrow A(i,i) < A(u,u).$$

The notion of an ESS pervades the entire literature on biological game theory. See, for example, Smith [1982] and Hofbauer-Sigmund [1988]. This notion is about to invade the field of economics as well—see, for example, Selten [1980]. We also encountered an ESS in Subsection 4.5.

5.3   Equilibrium Theorems of Mathematical Economics

In present-day mathematical economics, the *description* of the actual economic situation in an economic unit (for example, in a nation) plays a subordinate role. Mathematical economists try to confront economic reality with some mathematical *model* in order to understand the system comprehensively. They have today a huge arsenal of possible models for this purpose. The particular model that is chosen for a given problem depends more on the situation than on the problem itself. For example, the problem may be applied to the nation as a whole, or it may be applied only to a small region. In each case, the purpose of collecting concrete data is to adjust some of the parameters of the model and thus enable it to give good and realistic prognoses. Most of the models that presently exist in mathematical economics make use of the following kinds of data:

> a list of economic agents (such as producers, consumers etc.)
> a list of commodities that play a role in the economy under review
> price lists for these commodities, and for each such price list, a set of particularly preferred actions, one set for every commodity.

Each of the actions is also a list that specifies a positive or negative quantity for each commodity. Negative values stand for consumption, and positive values stand for production. An *equilibrium* consists of a price list and one action for each commodity, such that this action is among the preferred ones for the given price list for each commodity.

Once again the question to ask is whether an equilibrium must exist for each model. There are positive answers to this question, and these answers are precisely the equilibrium theorems of mathematical economics. They are usually proved by skillful applications of either Brouwer's or Kakutani's fixed-point theorem. See Section 5 in Chapter V.

The latter equilibrium theorems have one feature in common with Nash's equilibrium theorem for noncooperative *n*-person games: they deal with a finite number of players (agents), and the mathematical notions of convexity and compactness play a fundamental role in their formulations and proofs. There are, however, some differences as well. The possible actions of an economic agent are not simply mixed strategies made from a finite list of pure strategies; they are taken from a much more complicated domain of possibilities. Further, strategies are evaluated not simply by monetary payoffs, but by much more sophisticated preferences. The noncooperative *n*-person games can, however, be incorporated into mathematical economics as very special models.

We do not have enough space here to enter into further details. Readers with sufficient background in mathematics may profit from the books of Debreu [1959], Hildenbrand-Hildenbrand [1975], Cassels [1981], and König-Neumann [1985], and may appreciate the encyclopedic treatment given in Aubin [1979].

Some objections have been raised against the mathematical theory of economic equilibria. The most important ones are the following:

(1) The theorems that are provided by these theories are only existence theorems: they tell us only that an equilibrium exists, not where it lies or how to find it. Even when they do allow us to compute equilibria, they do not tell us how to realize these equilibria in a real-life situation because they do not provide access to the institutions that have the political power to push them through.

(2) The theory tells us very little about what happens in the absence of an equilibrium.

The second objection may be countered by referring to the mathematical theory of business cycles (see, for example, Rosenmüller [1972]). For more on the first objection see, for example, Kornai [1971] and Kötter [1982].

Chapter V · *Topology*

IN THIS SHORT chapter we introduce the reader to some of the basic notions of an important branch of mathematics known as *topology*, which provides exact mathematical models for such intuitive notions as

> continuum   continuity   deformation.

During the course of this chapter we shall also enrich our imaginations by considering some of the other notions of topology such as

> connectedness   mainfold   orientation   dimension   compactness.

In their attempts to provide precise meanings for these notions, mathematicians were often led astray by too naive a trust in their intuition, and it is only during the twentieth century that the topological foundations for these notions have been properly understood. As topology stands today, the details of these notions are not very complicated, but a precise and complete treatment of the subject would require too much space and would be too abstract to be appropriate for this book. We shall confine ourselves to a brief survey in which we shall often rely on our intuition to give meaning to the concept being discussed. Our main goal in this chapter will be to gain an intuitive acquaintance with some of the fundamental notions of topology and the way they differ from one another. You might consider this chapter as a little exhibition hall in which some particularly sophisticated examples of mathematical constructions are on display. For further reading on this subject we refer the reader to the chapters on topology in Coxeter [1969].

§1   *Topological Spaces and Continuous Mappings*

Consider the sketch shown in Figure V.1.1. With every point x on line g we associate the point of intersection y of the semicircle H and the line segment that joins O and x. We see that as x varies on the infinite line g, the point y varies on the semicircle.

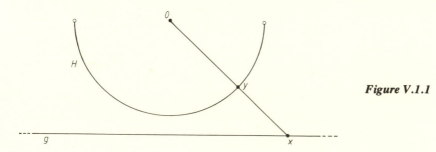

Figure V.1.1

We now make two intuitively obvious observations about the motion of the point y as x varies.

(a) If x varies by only a little, then y varies by only a little. Conversely, we can make the variation in x small by making the variation in y small enough.

(b) As x approaches infinity either to the left or the right of the line g, the point y approaches the corresponding endpoint of H. Note that y never actually reaches the endpoint of H.

Observation (a) can be made more precise by using the notion of a neighborhood. You may think of a neighborhood of a point in this context as being a small arc that has the point at its center.

(a*) Suppose that a point y_0 in H corresponds with a point x_0 in g. Then the following two conditions must hold (see Figure V.1.2):

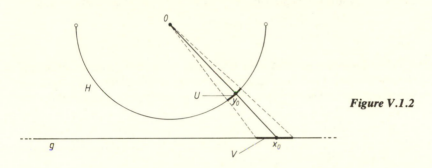

Figure V.1.2

(i) Given any neighborhood V of y_0, there is a neighborhood U of x_0 such that for every point x in U, the point y in H corresponding to x must lie in V.

(ii) Given any neighborhood U of x_0, there is a neighborhood V of y_0 such that for every point y in V, the point x in g corresponding to y must lie in U.

The line and semi-circle in Figure V.1.2 are examples of *topological spaces*, and the mapping we have described between these spaces is a *homeomorphism*. A function from one space onto another is a homeomorphism if it is continuous and one-one, and its inverse function is also continuous. When it is possible to find a homeomorphism from one topological space to another, we say that the two spaces are *homeomorphic*, or *topologically equivalent*. In the discussion that follows we shall consider four basic notions of topology (Kelley [1955] is a standard reference for mathematicians):

(1) *Topological space.* As we have said, the semicircle and line that we considered in the introduction to this section are examples of topological spaces. In general, a topological space consists of a set X (whose elements

are called the *points* of the space) and a *neighborhood structure*. A neighborhood structure associates to each point x of the space X a family of subsets of X (that we call the *neighborhoods* of x) such that the following two conditions hold:

(a) Every neighborhood of a point x contains the point x.

(b) Given any two neighborhoods U and U' of a point x, it is possible to find a third neighborhood $\bar{U}$ of x such that $\bar{U}$ is included in both U and U'. In other words, it is possible to find a neighborhood $\bar{U}$ of x such that

$$\bar{U} \subseteq U \cap U'.$$

See Figure V.1.3.

Most important topological spaces also satisfy a further condition known as the *Hausdorff separation axiom* and are therefore called *Hausdorff spaces*. The Hausdorff axiom says that whenever x and y are distinct points of the space X, it is possible to find a neighborhood U of x and a neighborhood V of y such that U and V are disjoint. In other words,

$$U \cap V = \emptyset.$$

See Figure V.1.4. The term Hausdorff space honors the mathematician, poet, and essayist Felix Hausdorff (1862–1942, pseudonym Paul Mongré). Hausdorff was a professor emeritus in Bonn in 1942 when he committed suicide in order to avoid deportation.

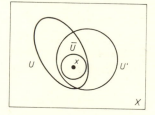

Figure V.1.3

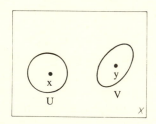

Figure V.1.4

(2) *Continuous function.* A function f from a topological space X to a topological space Y is said to be *continuous* at a point x_0 of X if for every neighborhood V of the point $y_0 = f(x_0)$ there exists a neighborhood U of x_0 such that $f(x)$ belongs to V whenever x belongs to U.

It is interesting to note that in spite of the enormous impact that the definition of continuity has in calculus, the definition depends only on the simple topological structure that we described in (1). The possibility of reducing the world of continua to a study of this simple structure was one of the exciting discoveries of the nineteenth century. If a function f from a topological space X to a topological space Y is continuous at every point of X, then we say that f is a *continuous function* from X to Y.

Before we leave this topic we shall take one more look at the definition of continuity and observe how it uses the *quantifiers "for every"* and *"there exists."* The universal quantifier ∀ means "for every," "for all," "for each," or "whenever," depending upon the way the sentence is phrased, and the existential quantifier ∃ says "there exists," "there is," "it is possible to find," or "for some." Now look at the definition of continuity again.

> A function f from a topological space X to a topological space Y is said to be continuous at a point x_0 of X if *for every* neighborhood V of the point $y_0 = f(x_0)$ *there exists* a neighborhood U of x_0 such that $f(x)$ belongs to V *whenever* x belongs to U.

Note that even the symbols, f, X, Y, and x_0 are preceded by a silent quantifier ∀ in this definition. Stated more fully (but more clumsily), the definition reads:

> *For every* topological space X and *every* topological space Y and *every* function f from X to Y and *every* point x_0 in X, the function f is said to be continuous at x_0 if . . .

We refrain from a systematic use of the symbols ∀ and ∃ in this book. For the professional mathematician they are daily shorthand. They—plus some other logical and mathematical symbols—are sufficient to formulate every statement in mathematics.

(3) *Homeomorphism.* As we have said, a homeomorphism from a topological space X to a topological space Y is a one-one continuous function from X onto Y such that the inverse function f^{-1} of f is continuous from Y to X. Recall that the condition that f is *one-one* says that if y is any element of the space Y, then there cannot be more than one element x in X for which $f(x) = y$. Furthermore, the condition that f is *"onto Y"* says that if y is any element of the space Y, then there must be at least one element x in X such that $f(x) = y$. Therefore, if f is one-one and also onto the space Y, then for each element y in Y there is *exactly one* element x in X such that $f(x) = y$. This unique element x is called $f^{-1}(y)$. Note that if f is a homeomorphism from X to Y, then f^{-1} is a homeomorphism from Y to X.

(4) *Topological Equivalence.* As we have said, two topological spaces X and Y are said to be *topologically equivalent* or *homeomorphic* if it is possible to find a homeomorphism from X to Y; or, equivalently, if it is possible to find a homeomorphism from Y to X. If two topological spaces X and Y are homeomorphic, then they have precisely the same topological properties, and for practical purposes we may think of them as being merely different names for the same space. To quote John Kelley: *A topologist is a man who doesn't know the difference between a doughnut and a coffee cup.*

By this means we are able to simplify enormously the problem of "classifying" all topological spaces: in a class of equivalent spaces we have to count only one of them. The problem of listing all possible distinct topological spaces without homeomorphic duplications is called the *classification problem for topological spaces* and represents a type of problem that arises all over mathematics (see also Chapter 1, §3). Whenever we have a mathematical concept we ask ourselves whether we can find all of the mathematical systems that fit this concept, and draw up a list of all these systems without repetitions. In a manner of speaking, we have completed our study of the concept as soon as we have drawn up such a list. For the concept "topological space" the classification problem is far from being solved nowadays. A complete list has been found only for some very restricted subclasses of topological spaces. See Section 3.

Sometimes our interest in a given topological space goes beyond the strictly topological structure of the space. When the actual construction of a given space is of interest in its own right, we may want to distinguish between two spaces even though they may be homeomorphic. For example, Figure V.1.5 shows an important topological space called the *Cantor discontinuum*, which is obtained by removing the middle third of a given interval, then removing the middle third of each of the two intervals that remain, then removing the middle third of each of the four intervals that remain, and continuing this process to infinity. Georg Cantor (1845–1918), who founded the theory of sets, discovered his discontinuum in about 1883 (for a certain priority of DuBois-Reymond, see Felscher [1978/79]).

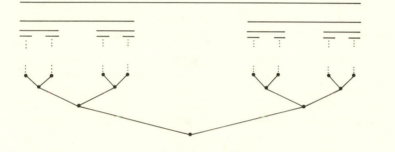

Figure V.1.5

Figure V.1.5 shows how the points of the Cantor discontinuum can be matched with the rising paths in the infinite binary tree that is shown below it. These paths can be identified, in turn, with infinite sequences of commands "go left" and "go right." If we now replace the commands "go left" and "go right" with the numbers 1 and 0, respectively, then we have matched the Cantor discontinuum with the set of all possible 0-1-sequences like 01101001 . . . The Cantor discontinuum is a topological space in a natural way because it is part of the number line with the usual neighborhood structure given there. On the other hand, the space of all infinite 0-1-sequences can be *made* into a topological space by assigning a collection of neighborhoods to each sequence. In fact, we assign a whole sequence of neighborhoods to each sequence where for each n, the nth neighborhood is the collection of all sequences that coincide with the given sequence up to the nth position. It is not hard to see that this topological space of 0-1-sequences is homeomorphic to the Cantor discontinuum. However, both Cantor's discontinuum and the space of 0-1-sequences are profoundly important mathematical constructions, and each is of interest in its own right.

Cantor's discontinuum plays a major role in many branches of mathematics and is the source of many interesting examples. One of these is Anatole Beck's paradox of *the hare and the tortoise.* In the classical fable of Babrios, the hare is clearly the faster animal, but he goes to sleep at some point in the road, and by the time he wakes up, he is too late to win the race. However, the fable according to Anatole Beck is much more subtle. In this version of the story, the hare is not allowed to sleep for a positive amount of time at any point in the road. Furthermore, at each point of the road, the speed of the hare is twice the speed of the tortoise. In spite of this, the hare is made to accumulate "sleeping time" in the Cantor discontinuum as he runs. Even though the tortoise also has a zero speed at each point of the Cantor discontinuum, he does not "sleep" there, and so once again the hare loses the race. If we want to, we can make the hare take twice as much time as the tortoise needs. See Figure V.1.6.

Position

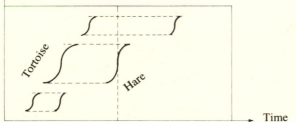

Time

Figure V.1.6

§2   *Curves and Knots*

Although curves are relatively simple examples of topological spaces, they give rise to extremely complicated phenomena, and the study of curves is very sophisticated. The notion of a curve is based on the following two simple examples:

The *unit interval* [0,1] contains all real numbers x for which $0 \le x \le 1$.

The *unit circle* C is the set of all ordered pairs (x,y) of real numbers for which $x^2 + y^2 = 1$.

See Figure V.2.1. Both [0,1] and C can be made into topological spaces by defining their neighborhoods in a natural way.

We can now give the general definition of a curve. A *curve* in a topological space X is a continuous function from the interval [0,1] into X. To get an idea of what a curve really

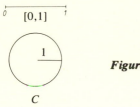

Figure V.2.1

is, you should think of a curve f in a space X as describing a *journey* in the space. We begin the journey at the point $f(0)$, and at each "time" t in the interval $[0,1]$ we arrive at the point $f(t)$. The journey ends at the point $f(1)$. In fact, if f is a curve in a space X, then the point $f(0)$ is called the *starting point* of f and the point $f(1)$ is called the *endpoint* of f. If the curve begins and ends at the same point, in other words if $f(0) = f(1)$, then we say that the curve f is *closed*. We can also think of a closed curve in a space X as being a continuous function from the circle C into X, provided that we think of C as beginning at some chosen point [usually the point $(1,0)$], and then winding once around the circle. It is important to distinguish between a curve f in a topological space X and the set of all points $f(t)$ where $0 \leq t \leq 1$. The latter set is the *range* or the *orbit* of the curve. For example, a curve that winds around a circle twice is not the same curve as one that winds around the circle once. But they have the same range.

If a given continuous function f from $[0,1]$ into a space X is a homeomorphism, then we say that the curve f is a *Jordan arc* in X. In the event that X is a Hausdorff space, then a curve f will automatically be a homeomorphism if it is one-one, and so for all common topological spaces we can simply define a Jordan arc to be a curve that does not pass through the same point twice. If f is a closed curve that is one-one on $[0,1]$ except for $f(0) = f(1)$, then we say that f is a *Jordan curve*. A Jordan curve can be seen as a homeomorphism from the circle C onto a subset of the given space. Jordan curves and arcs are named after Camille Jordan (1838–1922), not to be confused with the theoretical physicist Pascual Jordan (1902–1980) after whom Jordan algebras are named.

Before we state the most important properties of Jordan curves, we need to introduce the topological notion of connectedness. The kind of connectedness to which we shall refer here is usually known as *pathwise connectedness*. There is a more general notion of connectedness with which we shall not be concerned. We say that a topological space X is (pathwise) connected if for every two points x and y in X there is a curve that starts at x and ends at y. Not all topological spaces are connected. An example of an extremely disconnected space is the Cantor discontinuum, because no two distinct points of the Cantor discontinuum can be joined by a curve. Any curve we would try to draw would need to jump over some of the missing "middle thirds."

One of the most important theorems about Jordan curves concerns Jordan curves in a plane. The *Jordan curve theorem* says that if f is a Jordan curve in a plane E, and if K is the range of f, then the set $E \setminus K$ of points of E that do not lie on K is not connected, but this set is the union of two sets that are connected. One of these two sets is bounded and is called the *interior* of K, and the other is unbounded and is called the *exterior* of K. See Figure V.2.2.

The Jordan curve theorem is a typical example of the historical development of topology. Before Jordan, only special kinds of Jordan curves had been considered, such as "piecewise straight" or "smooth" curves, and many people regarded the theorem as

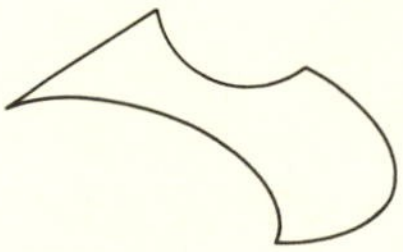

Figure V.2.2

trivial. It was Camille Jordan who first realized that the general definition of a curve allows examples that are so intricate that the theorem is anything but trivial. The proof he gave was very complicated, and even today a proof of the Jordan curve theorem can be given only after some elaborate preparation. See Schmidt [1923] and Moise [1977]. As deep as it is, the Jordan curve theorem can be even further refined. The interior of a Jordan curve is homeomorphic to a disc (without its boundary) and the exterior of the curve is homeomorphic to the plane with a disc (together with its boundary) removed. This form of the theorem is known as the *Schoenflies theorem*.

There is an anologue of Jordan's curve theorem in higher dimensions; for example, a spherical surface splits the three-dimensional space into two connected domains, its interior and exterior. But, as we shall see by an example—Alexander's horned sphere— the Schoenflies extension of Jordan's theorem does not carry over in the same fashion.

Jordan curves in three-dimensional space can be much more complicated than those in the plane. For example, the "clover-leaf" or "trefoil" shaped Jordan curves shown in Figure V.2.3 are called *knots*. Two knots are said to be *equivalent* if it is possible to

Figure V.2.3

deform one of them continuously into the other. In order to state this definition precisely we need to make us of the concept of a *deformation*. Roughly speaking, a curve f may be continuously deformed into g if we can find a family of curves f_t where $0 \le t \le 1$ such that $f_0 = f$ and $f_1 = g$, and such that the curve f_t "varies continuously" as t increases from 0 to 1. A deformation of a circle in three-dimensional space is depicted in Figure V.2.4. If a knot can be deformed into a curve that is "piecewise straight" or *polygonal*, then we say that the knot is *tame*. Figure V.2.5 shows that the clover-leaf knots are tame.

Figure V.2.4   *Figure V.2.5*

There are knots that are not tame and these are called *wild knots*. An example of a wild knot is given in Figure V.2.6.

Figure V.2.6

Let us throw a glance, in passing, at another wild figure (though not a curve): the topological space that is known as *Antoine's necklace*. This space is obtained by a process that is similar to the one we used to construct Cantor's discontinuum. In this case we successively eliminate parts from a solid ring. See Figure V.2.7.

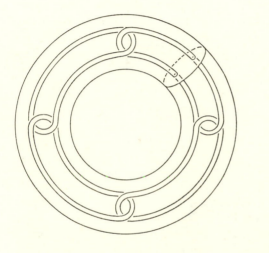

Figure V.2.7

Let us imagine that the original ring was carved out of a block of wood by the youngest apprentice in a carver's workshop. A more experienced apprentice then carves a chain of four links out of this ring. A still more senior apprentice carves a chain of four links out of each of the four links that the second man had made. This process can be continued indefinitely at least by mathematicians (even if wood carvers would eventually have to give up). The space that remains is Antoine's necklace. It was discovered by Louis August Antoine (1888–1971), who was blinded during the First World War. The intricacy of Antoine's construction is said to have provoked Erhard Schmidt (1876–1959) to say: "Only a blind man could have come up with something like this."

We shall return now to the study of tame knots. For these, we may pose the following *classification problem: Give a complete list of all the tame knots. In other words, give a list of tame knots such that every tame knot can be deformed into precisely one of the knots in the list.* Preliminary work that goes far back into the nineteenth century has culminated in the beginnings of such a list. Figure V.2.8 shows the list as it appears in Alexander-Briggs [1927] and Reidemeister [1932].

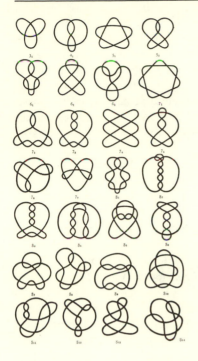

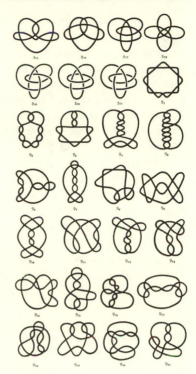

Figure V.2.8

The table shown in Figure V.2.8 lists all plane projections of tame knots for which at most nine crossings are needed. In the case of some of these projections, the spatial counterimage can immediately be reconstructed by overcrossings and undercrossings. In the other cases, the projection has two spatial counterimages that are mirror images of each other. Each of these two mirror images can be obtained from the other by interchanging the overcrossings and undercrossings. We can always obtain two mirror images this way, but in some cases it may be possible to deform each mirror image into the other. When this happens we say that the knot is *amphicheiral*. The clover-leaf knots are not amphicheiral.

In Figure V.2.8, the knots are given in order of increasing degree of complication, which may be defined as the minimum number of crossings that are needed in a plane projection of the knot. For example, the knots 8_1–8_{21} form the complete list of all tame knots with eight crossings. The list in Figure V.2.8 goes up to the knots with nine crossings. At present, knot theorists are trying to classify all the tame knots with up to eleven crossings.

There are six famous ornamental knots by Dürer (1505)—see Figure V.2.9—who, in turn, had been stimulated by the knot designs of Leonardo da Vinci; I do not know whether their topological nature has been investigated yet. Incidentally, knotted chain molecules play an important role in chemistry (Boeckmann-Schill [1974]).

So far, we have concentrated our attention on Jordan curves because these represent the simplest case and more results have been obtained about them. However, there are also interesting questions about more general curves, curves that may pass through a given point many times. Perhaps the simplest of all curves is a constant curve f that stays at the point $f(0)$ all the time; in other words, $f(t) = f(0)$ for every t. Figure V.2.10 depicts a constant curve and another curve that passes twice through a point on its orbit.

One of the most sophisticated examples of a curve is the *Peano curve*, which passes through every point in a square infinitely many times. This curve was discovered by Giuseppe Peano (1858–1932). Figure V.2.11 shows a variation of Peano's original construction that was given by Hilbert in 1891. This curve is the result of an infinite sequence of reshapings of a fairly simple starting curve.

This curve passes through every point of the *unit square* Q that contains all the points (x,y) for which $0 \leq x \leq 1$ and $0 \leq y \leq 1$. The square Q is thus represented as a continuous image of the unit interval $[0,1]$. A natural question that arose after Peano's discovery was whether Q could be represented as a Jordan arc or a Jordan curve. In other words, this question asks whether Q can be homeomorphic to either of the spaces $[0,1]$ and C.

The answer to both of these questions is no! These questions are quite easy to answer, but their higher-dimensional analogues are quite sophisticated. These questions and their higher-dimensional analogues were answered by Luitzen Egbertus Jan Brouwer (1881–1966) in his theorem on the *topological invariance of dimension*. For example, Brouwer's result tells us that the square Q cannot be homeomorphic to the three-dimensional cube. In proving his theorem, Brouwer laid the foundations of vast areas of modern topology. In 1928, Brouwer's theorem on invariance of dimension was provided with an extremely elegant proof by young Emanuel Sperner (1905–1980); See Sperner [1928]. The key to Sperner's proof is a combinatorial result which is nowadays called *Sperner's lemma*. We shall mention this lemma again in Section 5, in connection with another important theorem of Brouwer's, namely, his *fixed-point theorem*.

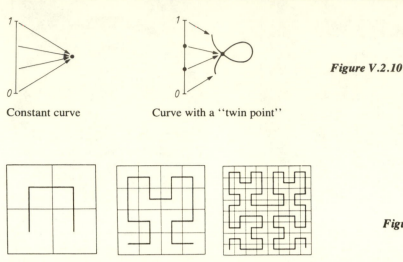

Constant curve   Curve with a "twin point"

Figure V.2.10

Figure V.2.11

We mention in passing that it was the same L.E.J. Brouwer who shook the foundations of mathematics at about the same time (around 1910) with his philosophy of *intuitionism*—logic without *tertium non datur*. Today, intuitionism is a recognized, though not widely adopted, branch of logic.

§3   *Surfaces*

Roughly speaking, a *surface* is any two-dimensional figure. We might consider a surface to be the range of a function that is continuous on a rectangle; or, in some contexts, we may consider a surface to be the function itself. Figure V.3.1 shows some examples of common surfaces. One surface that is not included in Figure V.3.1 is the most obvious one—the plane. In studying surfaces, we commonly distinguish between bounded surfaces and unbounded surfaces, and between surfaces with a boundary and surfaces without a boundary. Figure V.3.1 roughly illustrates these distinctions to some extent.

We shall be concerned mostly with bounded surfaces. Some of these, like the unit square, the unit disc, and the Möbius strip, will have boundaries; others, like the sphere and the torus, will not. The sphere and the torus are called closed surfaces. Surfaces are special kinds of topological spaces. Some of the surfaces shown in Figure V.3.1 are topologically equivalent. For example, you may deform a torus into a sphere with one "handle."

The clover-leaf tire in Figure V.3.1 is also equivalent to the torus, just as the clover-leaf knots in Section 2 were homeomorphic to the circle. Note that the fact that a given curve or surface is knotted cannot be felt by a topologist confined to a one- or two-dimensional life on the curve or surface, respectively: it is not an intrinsic topological property of the latter, but depends, instead, on how the curve or surface has been placed

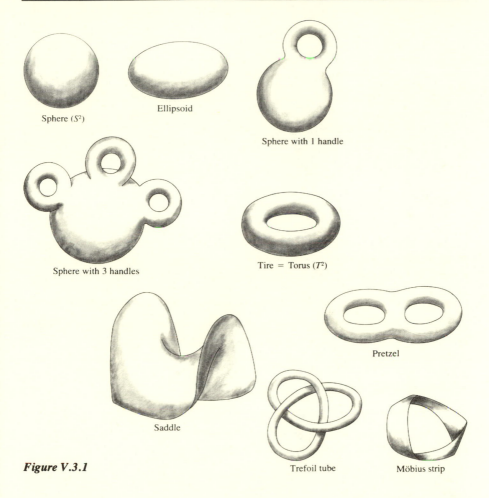

Sphere (S^2)

Ellipsoid

Sphere with 1 handle

Sphere with 3 handles

Tire = Torus (T^2)

Saddle

Pretzel

Figure V.3.1

Trefoil tube

Möbius strip

in space. Figure V.3.2 shows that a sphere with one point (the "north pole") removed is homeomorphic to the plane, and that the plane is also homeomorphic to a hemisphere without its rim.

In this section we shall be concerned with the problem of *classification of all closed surfaces*. This is the problem of listing all the closed surfaces in such a way that every closed surface is homeomorphic to precisely one surface in the list. In particular, no two surfaces in the list are topologically equivalent.

We shall now take a look at the way some of the common surfaces arise. The sphere of radius 1 can be seen as the solution set of the equation

(1) $x^2 + y^2 + z^2 = 1$

in the space R^3 that consists of all triples (x,y,z) of real numbers. Ellipsoids, paraboloids, and hyperboloids (the so-called *quadratic surfaces*) also appear in this way, and the study

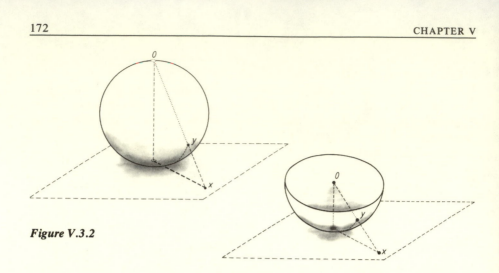

Figure V.3.2

of these particular surfaces is one of the features of classical analytical geometry. In general, the solution set of an equation of the form

(2) $f(x, y, z) = 0$

will be a surface if f is a properly behaved function. There is, however, an important special case of equations of this type, and this is an equation of the form

$$z - f(x, y) = 0,$$

which we can write in the form

$$z = f(x, y).$$

The solution set of an equation of the latter form is the graph of the function f of two variables. For example, if f_+ and f_- are defined as

$$f_+(x) = \sqrt{1 - (x^2 + y^2)}$$

and

$$f_-(x) = -\sqrt{1 - (x^2 + y^2)},$$

where the point (x, y) lies in the disc $x^2 + y^2 \leq 1$, then the graph of f_+ is the set of all points of the form

$$(x, y, f_+(x, y)) = \left(x, y, \ \sqrt{1 - (x^2 + y^2)} \ \right),$$

and this set is the upper hemisphere of the sphere (1). Similarly, the graph of f_- is the lower hemisphere of the sphere (1).

This example illustrates a general principle: If a surface can be represented as the solution set of an equation of type (2), then under reasonable conditions it may be partitioned into a number of pieces, each of which can be represented in the form

$$z = g(x, y).$$

In the case of the sphere, two pieces were needed, and the two functions g are the functions we called f_+ and f_-. The mechanism by which a surface may be represented in this way is an important theorem in calculus known as the *implicit function theorem*.

The method of describing surfaces as the solution sets of equations or as the graphs of functions is clearly rigorous mathematics, but there is another and most important device that allows us to manufacture new surfaces out of old ones. This method involves cutting surfaces, bending them, stretching them, and gluing their rims together. For example, we can manufacture a Möbius strip by deforming a rectangle and then pasting together two of its rims, as shown in Figure V.3.3.

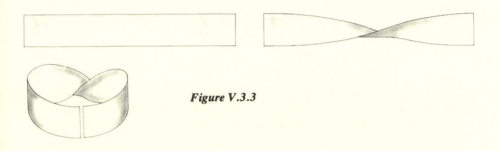

Figure V.3.3

Putting it quite simply, we might say that you can make a Möbius strip by opening your belt, twisting it, and then closing it again. Figure V.3.4 shows how to manufacture a torus out of a rectangle. Figure V.3.5 then shows how the torus may be converted into a pretzel surface by adding one handle.

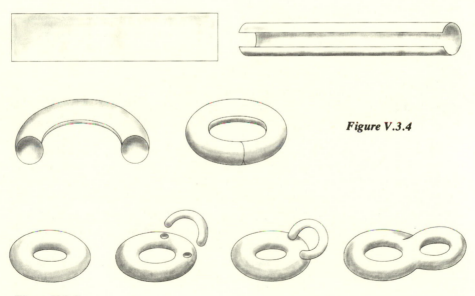

Figure V.3.4

Figure V.3.5

Although this process of cutting, bending, stretching and pasting does not look like rigorous mathematics, it can be made perfectly rigorous and is today a common technique employed by topologists, who refer to it as "surgery." For example, we can obtain some idea of the exact mathematical version of the construction of the Möbius strip shown in Figure V.3.3 by looking at this construction as follows: We start with a rectangle, and then *identify* every point of the left boundary with a point in the right boundary, as shown in Figure V.3.6.

Figure V.3.6

The shaded area becomes a typical neighborhood of the new single point that results when the points in the left boundary and right boundary are identified. This kind of description of the topologists' "surgery" can be used to turn it into rigorous mathematics.

The kind of method that we have just described for interpreting topological surgery has an additional advantage—it allows us to perform surgery that cannot be achieved physically in three-demensional space. For example, the famous *Klein bottle* (Klein [1882], p. 571) can be constructed mathematically as shown in Figure V.3.7. The only way we can show the Klein bottle in three-dimensional space is by cheating. We have to show the bottle with a self-penetration that is not supposed to be there.

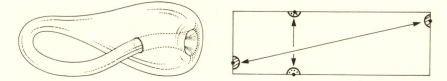

Figure V.3.7

It is interesting to note that the surface obtained from a hemisphere by identification of opposite points as shown in Figure V.3.8 would yield a surface topologically equivalent to the Möbius strip if we were to cut off the parts that are shaded in Figure V.3.8. If we leave the surface intact, we can imagine ourselves going to infinity along a line and then returning from the other side after having passed the "point at infinity." See Figure V.3.9. "Points at infinity" are a feature of the projective geometry that we discussed in Chapter I, and for this reason, our surface is also called the *topological projective plane*. If we imagine a crosswise sewing of seams,[1] then we may speak of the *sphere with one cross-cap*.

[1] That's how a bachelor mends his trousers.

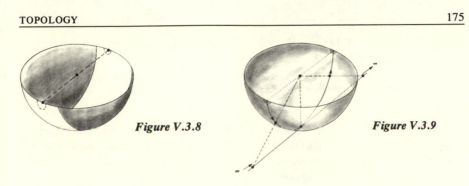

Figure V.3.8   *Figure V.3.9*

To see the effect of such crosswise identifications, look at the Möbius strip in Figure V.3.10. The little man becomes his own antipode if he walks around suitably (gravity-free, of course). Now look once more at the topological projective plane as shown in

Figure V.3.10

Figure V.3.11. A right-oriented triangle passes through infinity and returns as a left-oriented triangle. For this reason, the Möbius strip and the topological projective plane are said to be *non-orientable surfaces*. Using the language that we have developed in this discussion, we can now present the solution of the classification problem for closed surfaces:

(1) Every orientable closed surface is topologically equivalent to precisely one

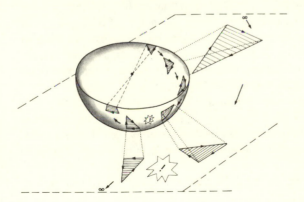

Figure V.3.11

surface of the form K_g , where K_g stands for a sphere with g handles. The number $g = 0,1 \ldots$ is called the *genus* of the surface.

(2) Every non-orientable closed surface is topologically equivalent to precisely one surface of the form $\bar{K}_h$, where $\bar{K}_h$ stands for a sphere with h cross-caps.

These statements remain true even if we redefine the concept of the "surface," combining all the manufacturing devices that we have mentioned in this section, and rename the new concept a "two-dimensional manifold," or, more briefly, a "2-manifold." We shall now present the essentials of this definition.

A disc with its boundary omitted is called an open disc and is shown in Figure V.3.12. Suppose that X is a topological space. A homeomorphism of an open disc onto a neighborhood of a point x of X will be called a *neighborhood map* of x. If every point x of the space X possesses such a neighborhood map, then X is said to be *two-dimensional* (note that Brouwer's dimension invariance theorem guarantees that X cannot have two different dimensions). For example, consider a globe of the earth. We can obtain a neighborhood map of any point on the globe by opening an atlas and finding a flat map that contains the given point. Don't mind the fact that the maps in your atlas are rectangles; this is no problem because open discs and open rectangles are all topologically equivalent.

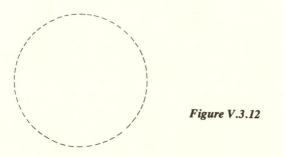

Figure V.3.12

With this example in mind, we call any choice of neighborhood maps for the points of a manifold, an *atlas* for that manifold. A two-dimensional mainfold is said to be *closed* if it possesses a finite atlas and is pathwise connected. The notion of orientability can also be defined in an exact manner, and we can therefore formulate the classification problem for closed surfaces in terms of two-dimensional manifolds. A proof that the classification we have described for surfaces also applies to two-dimensional manifolds may be found in the classical work by Seifert-Threlfall [1934].

The notion of a manifold can also be extended to dimensions other than two, and enhanced by additional structural conditions. For example, we may speak of a "smooth manifold." This topic dominates a large part of modern topology, and is also a major topic in the field known as *differential geometry*. For its role in the theory of general relativity see Sachs-Wu [1977].

All the sketches and constructions of manifolds that have been given up until now in this section have led to examples of "tame" mainfolds. For example, any manifold that can be obtained by pasting together a finite number of triangles in space is a tame manifold. However, mathematicians sometimes have good reason to describe some "wild" surfaces. The most famous wild surface is the *Alexander horned sphere* (Alexander [1924]). The Alexander horned sphere is topologically equivalent to a sphere because it can be obtained by a set of sophisticated deformations in which parts of the "horns" approach each other without touching, as in Figure V.3.13. When this process is continued indefinitely, the Alexander horned sphere results. See Figure V.3.14.

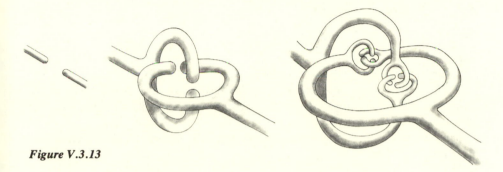

Figure V.3.13

The important fact about the Alexander horned sphere is that it is knotted. If you tie a string around an ordinary sphere, you can always pull it free. However, this is not the case with the Alexander horned sphere, because it is impossible to pull a string through the locked horns. The Alexander horned sphere therefore shows us that just as knotted curves can be homeomorphic to unknotted curves, so can a knotted surface be homeomorphic to an unknotted surface. It also shows, as announced above, that the exterior of

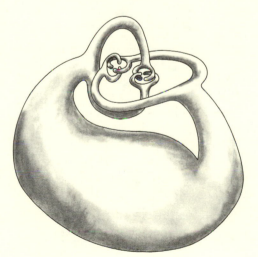

Figure V.3.14

a surface in three-space that is topologically equivalent to a sphere, may be topologically inequivalent to the exterior of a sphere, so that an anologue of the two-dimensional Schoenflies theorem does not hold. A precise mathematical formulation of the notion of a *knotted surface*, and the ability to "pull a string free," can be given in terms of what we call the *fundamental group* of a topological space. The fact that the Alexander horned sphere is knotted depends upon the fundamental group of the space that remains when the Alexander horned sphere is removed from three-dimensional space. We shall encounter fundamental groups in Subsection 4.2.

§4   *Curves on Surfaces*

It is sometimes possible to obtain information about a complicated mathematical shape by inserting a simpler object as a sort of probe, letting the simpler object tour the complicated one, and then listening to the simpler one as it tells us about its experiences during the voyage. For example, in the last section, we were able to recognize the non-orientability of the Möbius strip or the topological projective plane by letting a small oriented triangle travel about on it.

In this section we shall obtain some information about surfaces by placing curves on them and letting them vary.

4.1   *Euler's Polyhedral Formula*

We begin by looking at the five Platonic solids and reading some numbers from their shapes. For each solid we define the numbers f, e, and p as follows: f = the number of faces; e = the number of edges; p = the number of vertices. Figure V.4.1 illustrates these solids and the values of $f, e,$ and p for each one.

Name	Figure	f	e	p	$f - e + p$
Tetrahedron		4	6	4	$4 - 6 + 4 = 2$
Cube		6	12	8	$6 - 12 + 8 = 2$
Octahedron		8	12	6	$8 - 12 + 6 = 2$
Dodecahedron		12	30	20	$12 - 30 + 20 = 2$
Icosahedron	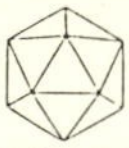	20	30	12	$20 - 30 + 12 = 2$

Figure V.4.1

Looking at Figure V.4.1 we make a startling observation: the *alternating sum* $f - e + p$ always has the same value, 2. The table shown in Figure V.4.2 demonstrates that this is not a peculiarity of the Platonic solids at all. Notice that the Platonic solids are all topologically equivalent to a sphere. The phenomenon arises whenever we draw a network of lines on a sphere, starting with one single point and then carrying out some or all of the following three operations a finite number of times in any order, allowing repetitions. Compare Fig. V.4.2.

Step no.	Figure	f	e	p	$f - e + p$
0		1	0	0	$1 - 0 + 0 = 1$
1		1	0	1	$1 - 0 + 1 = 2$
2		1	1	2	$1 - 1 + 2 = 2$
3		1	2	3	$1 - 2 + 3 = 2$
4		1	3	4	$1 - 3 + 4 = 2$
5		2	4	4	$2 - 4 + 4 = 2$

Figure V.4.2

    (a) Insert a point into an existing edge.
    (b) Draw a new edge that runs form an existing vertex to a new vertex without intersecting with any existing edge.
    (c) Draw a new edge that runs from one existing vertex to another without intersecting with any existing edge.

We see that in every case the alternating sum $f - e + p$ is equal to 2.

The natural question to ask is, what makes the number $f - e + p$ remain constant? To answer this question we should notice that each of the operations (a), (b), and (c) increases two of the terms in the sum $f - e + p$ that have opposite signs, and therefore leaves the sum constant.

For the torus shown in Figure V.4.3 we find that $f = 1$, $e = 2$, and $p = 1$; thus $f - e + p = 0$. However, this is because the shape of the torus is fundamentally different from that of the sphere, and the edges in the figure are able to sense this difference. It is not

Figure V.4.3

hard to show that the number $f - e + p$ has the value $2 - 2g$ whenever we draw a rich enough network of curves using the operation (a), (b), and (c) on a sphere with g handles. The network need only be rich enough to be able to sense all of the g handles. This is the case if all of the f faces are topologically equivalent to discs. For the torus considered in Figure V.4.3 we have $g = 1$, and so $f - e + p = 1 - 2 + 1 = 0$. We remark that all of the curves drawn in these figures are meant to be Jordan arcs or Jordan curves, and that our counting of f makes use of the Jordan curve theorem and the Schoenflies theorem.

4.2   The Fundamental Group

In this subsection we look at another way of using curves as "probes" on a surface. We begin by selecting a point on the surface, and then we draw closed curves that start and end at this point. We regard two such curves as being essentially the same (*homotopic*) if one of them can be deformed into the other within the surface.

To illustrate this idea, we consider the case in which the surface is an annulus. Figure V.4.4 demonstrates how two curves in the annulus can be deformed into one another if

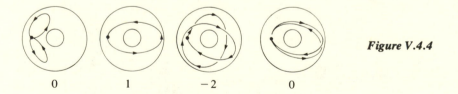

Figure V.4.4

      0   1   −2   0

and only if they have the same *winding number* with respect to the hole in the center of the annulus. Roughly speaking, this winding number of a curve is the number of times that the curve winds around the hole in a counterclockwise direction. If the curve winds around the hole clockwise, then its winding number is negative.

These curves can be combined by a natural process called *concatenation*. Roughly speaking, the concatenation of two curves is defined by running along the first curve and then running along the second curve. An important property of the winding numbers is that if f and g are two curves that have winding numbers m and n, respectively, then the concatenation $f \circ g$ of f and g has winding number $m + n$. In other words, the algebraic operation of concatenating curves entails the operation of addition in the system of integers. See Figure V.4.5.

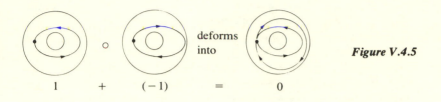

$$1 \qquad + \qquad (-1) \qquad = \qquad 0$$

Figure V.4.5

We see, then, that the process of concatenation turns the family of curves into a *group*. Strictly speaking, we are not looking at the curves themselves, but at the classes of curves that are homotopic to one another; in other words, the classes of curves that have the same winding number. The identity of this group is the family of curves that have winding number zero. Furthermore, if a given class of curves has winding number n, then the inverse of this class is the class of curves that have winding numbers $-n$. The group that is associated with a given surface in this way is called the *fundamental group* of the surface. The notion of a fundamental group was suggested by Poincaré [1895]. The notion of a fundamental group can be defined for arbitrary topological spaces, and leads to the following result:

> If two pathwise connected topological spaces are topologically equivalent, then their fundamental groups are the same.

Note that when we say that their fundamental groups are the same, we mean only that the groups can be matched in a one-one fashion in such a way that the algebraic combinations of elements match also. In other words, we mean that the groups are *isomorphic*.

Thus if two pathwise connected topological spaces have different fundamental groups, then they cannot be topologically equivalent. For example, the torus and the sphere do not have the same fundamental group, and, more generally, the fundamental groups show that a sphere with g_1 handles can never be topologically equivalent to a sphere with g_2 handles if $g_1 \neq g_2$.

Apart from the fundamental group of a given surface, we sometimes need to look at the fundamental group of the *exterior space*, the space that remains when the surface is removed. Roughly speaking, the exterior space is what remains if we carve out a solid figure bounded by the surface from the space in which the surface lies.

The fundamental group of a surface is related to the network of curves that we considered in Section 3 as follows:

> Every closed curve on the surface can be deformed inside the surface into a curve that consists entirely of edges from the network.

> Deformations of such curves can be performed by sliding them over faces in the network.

We conclude therefore that Euler's polyhedral formula is a statement about the fundamental group of the surface in question.

§5   *Compactness*

A form of the pigeonhole principle tells us that if you play the piano for an infinitely long time, then there must be at least one key that is pressed an infinite number of times. On the other hand, if you play the violin for an infinitely long time and play a succession of well-separated staccato notes, then (especially if you are a poor player) there need not be any note that is repeated. What we can say, however, is that there must be at least one note that is approximated infinitely often and arbitrarily closely. To understand why, we could subdivide the violin finger-board into a finite number of parts (like the finger-board of a guitar) and use the pigeonhole principle to find a region that is played infinitely often. Then we could subdivide this region into smaller pieces and use the pigeonhole principle again to find one of these smaller pieces that is played infinitely often. By continuing this process to infinity we could eventually find the point on the finger-board that we are looking for.

The existence of this point depends on the combination of two ideas:

The repeated application of the piegeonhole principle.

The existence of a point that lies in every one of the contracting sequence of regions on the finger-board that we have described.

These ideas apply just as well to two-dimensional regions. We could say, for example, that if you beat a drum infinitely often, then some point on the drum must be approximated infinitely often and infinitely closely. In fact, the principle applies in higher-dimensional spaces too. In general, topological spaces to which this sort of principle applies are said to be *compact*. The precise definition of a compact space is a follows:

Definition 5.1. A topological space X is said to be *compact* if it has the following *covering property*. If we associate to every point x in the space, a neighborhood U_x of x, then it is possible to find a finite number of points—call them $x_1, x_2, \ldots, x_n$—in X such that every point x in the space is contained in at least one of the neighborhoods U_{x_i}.

The definition of compactness is precisely what we need in order to apply the pigeonhole principle. The definition allows us to subdivide the space into arbitrarily small parts (using arbitrarily small neighborhoods U_x) and then to choose a finite number of these parts that "cover" the whole space.

One of the principles that we required in order to find the desired point in the preceding violin example is that a contracting sequence of regions on the board should have a point in common. A general form of this principle, known as *Cantor's intersection theorem*, holds in every compact topological space. To state this theorem we need the definition of a closed set and of a contracting sequence. A sequence $A_1, A_2, \ldots$ of sets is said to be *contracting* if $A_{n+1} \subseteq A_n$ for every n. A subset M of a topological space X is said to be *closed* if for every point x in the space, if x does *not* lie in M, then x has a neighborhood U_x that does not intersect with M. See Figure V.5.1. We can now state Cantor's intersection theorem

Theorem 5.2. Every contracting sequence of nonempty closed subsets of a compact topological space has at least one point in common. In other words, there is at least one point x in the space that belongs to every member of the given sequence.

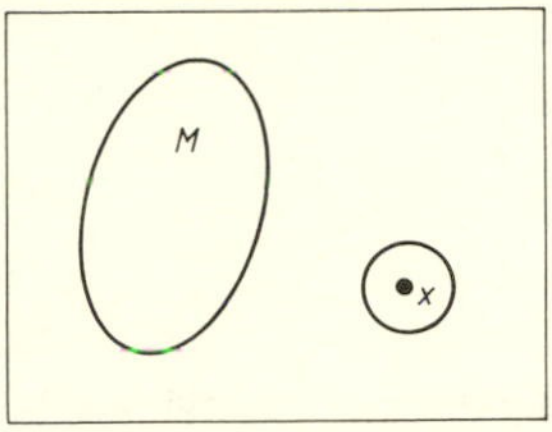

Figure V.5.1

Cantor's intersection theorem follows very simply from the definition of compactness.

An important theorem known as the Heine-Borel theorem tells us that every closed, bounded subset of the line $\mathbb{R}$, the plane $\mathbb{R}^2$, or the three-dimensional space $\mathbb{R}^3$ must be compact as a topological space. In general, every closed, bounded subset of Euclidean space must be compact. In particular, the following spaces are all compact: the violin string, the interval $[0,1]$, the membrane of a drum (which consists of a disc together with its boundary), a closed curve (and therefore, every knot), and a closed surface.

As we have said in Chapter III, the pigeonhole principle is an example of a pure existence theorem. It tells us that something exists (the pigeonhole with more than one pigeon in it) without telling us how to find it. Compactness plays a similar role, and in any mathematical theorem in which we have used compactness to prove existence, we have proved the existence without giving a constructive method of finding the object whose existence has been proved. We can think of compactness as being the right to use the pigeonhole principle infinitely often. So, for example, the Cantor intersection theorem tells us that a given contracting sequence of sets must have a point in common, but it does not tell us where this point is.

Another pure existence theorem that is extremely deep and that has a large number of important applications is *Brouwer's fixed-point theorem* (Brouwer [1911]), which we have already mentioned on a number of occasions. The statement of Brouwer's theorem is as follows:

Theorem 5.3. Brouwer's Fixed-Point Theorem. Suppose that X is a topological space that is topologically equivalent to a closed interval, a closed disc, a closed ball, or the analogue of any of these in higher-dimensional spaces. Then X has the following *fixed-point property*: Every continuous function T from X to X has at least one fixed point. In other words, if T is a continuous function from X to X, then it is possible to find a point x in X such that $T(x) = x$.

Among the continuous functions from a space X to X is the *identity mapping* T defined by $T(x) = x$ for every x. This function certainly has a fixed point, because every point of X is fixed. Another simple example is a constant function that sends every point x of the space X to a given point x_0. In this case the point x_0 is fixed. We should mention that Brouwer's theorem does not apply to all spaces, nor even to all compact spaces. For example, a rotation of the circle line by a given angle α is always a continuous function of the circle into itself, but (except for some special angles) it has no fixed point. The theorem also fails if the function T is not continuous. For example, the "puzzle map" shown in Figure V.5.2 has no fixed point if the endpoints of the subintervals are handled suitably.

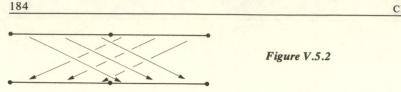

Figure V.5.2

For applications, we often make use of the fact that every compact, convex set satisfies the hypotheses of Brouwer's theorem. A subset S of the plane, three-dimensional space, or any Euclidean space is said to be *convex* if for every two points x and y in S, the line segment joining x to y is a subset of S. See Figure V.5.3.

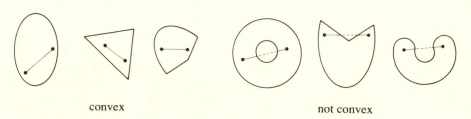

            convex   not convex

Figure V.5.3

We might restate this definition in the following visual fashion: S is convex if and only if any two of its points "see each other" within S. The notion of convexity was introduced by Herman Minkowski (1864–1909) specifically for use in pure mathematics, but this concept has turned out to be enormously useful for a host of applications—see, for example, Chapter IV. Those who apply mathematics to other disciplines usually see Brouwer's theorem in the following form:

Theorem 5.4. (Brouwer's Fixed-Point Theorem). Suppose that X is a convex, compact subset of Euclidean space. Then every continuous function from X to X has a fixed point.

The following generalization of Brouwer's theorem is even better tailored to many applications.

Theorem 5.5. (Kakutani's Fixed-Point Theorem). Suppose that X is a compact, convex subset of Euclidean space. Suppose that to every point x in X we associate a convex, compact subset $T(x)$ of X, and assume that this association is continuous in the following sense: If x' is near x, then $T(x')$ is near $T(x)$. Then there is at least one point x in X such that x belongs to $T(x)$. See Figure V.5.4. Kakutani's theorem can be derived from Brouwer's, which is, in turn, a special case of it. In the Brouwer form of the theorem, all the sets $T(x')$ are single points. See, for example, Burger [1959] and Franklin [1980].

Those who use these theorems for applications to other fields will often want to know how the fixed points may actually be found. To satisfy these requests, the modern proof of Brouwer's theorem using *Sperner's lemma* has been written as an algorithm. (See Scarf [1973].) We shall now present Sperner's lemma in some detail. Emanuel Sperner

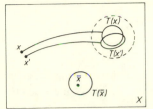

Figure V.5.4

(1905–1980) published it in 1928 as his key tool for a new proof of Brouwer's theorem on dimensional invariance (Sperner [1928]). It then turned out that his lemma was very well suited to provide a proof of the fixed-point theorem as well. Sperner's lemma deals with figures that we call *simplexes*. These are figures such as intervals, triangles, tetrahedra, and so on. Stated for triangles, Sperner's lemma can be put as follows:

> Suppose that the triangle is subdivided as shown in Figure V.5.5. (A subdivision of this form is usually called a *simplicial subdivision*.) Suppose that every vertex of every triangle in this subdivision is associated in some way with one of the vertices of the original triangle, and suppose that this association is performed in such a way that whenever a vertex of one of the smaller triangles lies on a side of the original triangle, this vertex is associated with one of the two endpoints of that side. Then there is at least one triangle in the subdivision whose vertices are associated to all three of the vertices of the original triangle.

In Figure V.5.5, we have shaded one of the five triangles that have the required property. Actually, Sperner's ingenious proof shows that the number of such triangles is always odd, and hence it is never zero.

Having stated this form of Sperner's lemma, we shall show how it may be used to derive the two-dimensional form of Brouwer's fixed-point theorem. We shall assume that the three vertices of the big triangle are numbered 0, 1, and 2. We shall say that a point x in the big triangle is *cool* to vertex number i if the distance from x to the side opposite i is not less than the distance from the point $T(x)$ to this side.

It may easily be shown that every point is cool to at least one vertex. In other words, the three domains

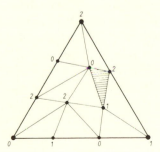

Figure V.5.5

$$\{x \mid x \text{ is cool to vertex } 0\}$$
$$\{x \mid x \text{ is cool to vertex } 1\}$$
$$\{x \mid x \text{ is cool to vertex } 2\}$$

cover the big triangle.

The idea of our proof is to find points that belong to all three of these domains. These "overall cool" points are precisely the fixed points of the function T. To find these points we begin by choosing a simplicial subdivision of the big triangle, and to every vertex x of a triangle in this subdivision we associate one of the vertices of the big triangle to which x is cool. It is easy to show that this may be done in such a way that it complies with the hypotheses of Sperner's lemma. We now choose a triangle of the subdivision whose vertices are associated to all three vertices of the big triangle (an "overall cool" triangle). Next we subdivide the big simplex into much smaller triangles. Applying Sperner's lemma again we obtain a smaller overall cool triangle. Repeating this process over and over again, we obtain a sequence of smaller and smaller "overall cool" triangles; by a compactness argument, we get a subsequence of these which converges to an overall cool point, i.e., to a fixed point of the function T. See, for example, Jacobs [1983a].

There are several topological results that are related to Brouwer's fixed-point theorem. Among these are the following:

(1) *The Porcupine Theorem.* A porcupine (in other words, a sphere with quills) cannot lay all of its quills flat. At least one quill stands out vertically at the center of a ridge.

(2) *The Sandwich Theorem.* Every peanut butter and jelly sandwich can be cut in two by a plane (knife) in such a way that each of the two parts have the same amount of bread, the same amount of peanut butter, and the same amount of jelly.

(1) is sometimes also called the *Hairy Ball Theorem.*

In this section we have tried to convey only a very few of the wealth of results that depend upon the notion of compactness. Because of the importance of compactness, it is of interest that many kinds of noncompact topological spaces can be inserted into larger spaces that are compact. This process is known as *compactification*. Look at Figure V.5.6 (which we have seen before). This figure tells us that an infinitely long line is topologically equivalent to a semicircle *without* its endpoints. This association of points on a line to points on a semicircle suggests that by adding two additional points $-\infty$ and ∞ to the line, we can make it topologically equivalent to the compact semicircle that *contains* its endpoints. In order to have a topological space after we have added in the two new points, we have to describe the neighborhood of $-\infty$ and ∞. These have to correspond to neighborhoods of the endpoints of the semicircle, and are hinted at in

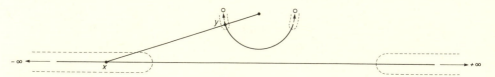

Figure V.5.6

Figure V.5.6. The space obtained in this way is known as a *two-point compactification of the line*.

Figure V.5.7 suggests how one may construct a *one-point compactification* of the line. Together with one new point ∞, the line becomes topologically equivalent to a full circle.

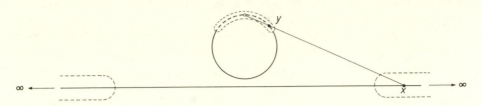

Figure V.5.7

This way of looking at the line represents the idea of "passing through infinity," which has a particular appeal to nonmathematicians. There is no mathematical theorem that says that one may "pass through infinity." Whether or not we can do this is simply an arrangement that is made by mathematicians, and we can choose whether or not we want to make this arrangement. The mathematical theorem says that if we want to make this arrangement, we can. This type of freedom in mathematics is sometimes threatening to nonmathematicians, who expect to find everything in mathematics cut and dried.

We conclude our brief remarks about compactification with Figure V.5.8, which sug-

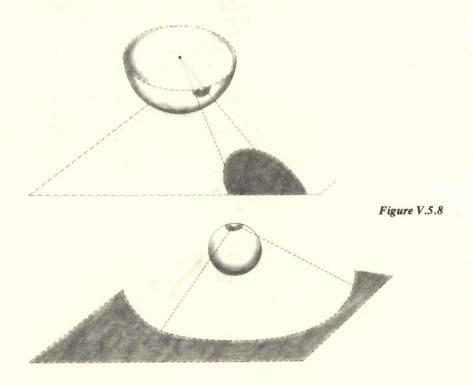

Figure V.5.8

gests how a plane may be compactified either by adding a "point at infinity" or by adding a "circle at infinity."

You might wish to ponder the notion of dimension in this context. The first of these two compactifications yields what is known as the *Riemann sphere* (after Bernhard Riemann [1826–1866]). The Riemann sphere allows us to look at a sphere as being the set of complex numbers with one new point ∞ added. The second of these two compactifications is the (topological) *projective plane*.

Mathematicians are sometimes inclined to use their terminology for jokes outside math. An example: A fictitious obituary for Bourbaki contains the words

Car Dieu est le compactifié . . . de l'univers (Groth. IV. 22)

suggesting God to be a manmade arrangement like the point ∞ in a compactification.

§6   *A Survey of Topology*

In the preceding sections we have acquainted the reader with a variety of typical notions and ideas in the field of topology. Topology germinated during the nineteenth century, and grew into a vast systematic theory during the twentieth. A community of mathematicians toiled in order to mold the concepts of topology and to make them precise so that they would lead to a fruitful development of the field. One of the most significant events in this development was the establishment of a link between topology and algebra. We hinted at the existence of this link when we discussed the notion of the fundamental group of a space. Group theory is only one of several branches of algebra that were originally called upon to serve topology, but that in turn have been and continue to be influenced by topology for their own development. Some of the developments in algebra that were motivated by topology have recently gained an enormous momentum in their own right. Algebraic disciplines of this type include *homological algebra* and *category theory*.

A gold mine of examples of topological spaces, some simple and some extremely sophisticated, can be found in Steen-Seebach [1970]. A solid introduction to topology is given in Kelley [1955] and Massey [1967]. For category theory see MacLane [1970].

Chapter VI · *Dynamics*

GENERALLY SPEAKING, *dynamics* is the theory of changing, while *statics* is the study of a single unchanging situation. Both the physicist and the mathematician look upon the situations that may occur as possible *states of a system*, and the system is thought of as being the collection of all possible states, together with the mathematical and physical laws according to which the states change. Even as the individual situations of states of the system change, the system itself, as a whole, is presumed to remain constant. Thus in both physics and mathematics, we may say that dynamics is the theory of changing states in a given system. Since "change by naught" = "no change," statics appears as a special case of dynamics.

If X denotes the set of all possible states of a given system, then the law according to which the states change is described by a mapping T from X into itself. The function T associates to every state x in the system X, its consecutive state $T(x)$. We may interpret this notation by saying that if x is the state of the system at any given instant, then one unit of time later, the state of the system will be $T(x)$. In other words, the mapping

$$T : X \rightarrow X$$
$$x \rightarrow T(x)$$

associates to each state x of the system, the state of the system one unit of time later. After a second unit of time has passed, the state of the system will be $T(T(x))$, which we also write as $T^2(x)$. Continuing in this fashion we obtain the sequence x, $T(x)$, $T^2(x)$, $T^3(x)$, . . . of all states through which the system passes, starting with the state x. This sequence is called the *orbit* of the state x. We shall also define $T^0(x) = x$. Thus for every nonnegative integer t, the symbol $T^t(x)$ denotes the state of the system after t units of time. Occasionally we write "x at time t" instead of $T^t(x)$. In the event that $T(x) = x$, the orbit of x is the constant sequence $x,x,x, . . .$, and we say that x is a *fixed point* of T. The theory of statics is the study of orbits of this type.

The set X and the mapping $T : X \rightarrow X$ form what mathematicians call the *dynamical system* (X,T). We can now describe the mathematical theory of dynamics as the theory

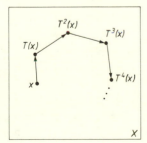

Figure VI.0.1

of the behavior of the orbits of the states in a dynamic system. Note that in the dynamical systems we are considering here, we have made two assumptions:

(1) Every time t is understood to be an integer multiple of a given unit of time. The system varies by changing "discretely" from state to state in much the same way that a movie changes on the screen by changing from one still picture to another.

(2) We assume that the law governing the way states change does not vary as time passes.

We should mention before we go any further that a significant part of the mathematical theory of dynamics goes beyond these two limitations. Dynamical systems have also been studied in which the time parameter t varies continuously, and there are theories that allow the possibility that the law governing changes of states may vary with time. In this introductory presentation of dynamics we shall usually assume that conditions (1) and (2) hold. We shall, however, make a brief detour into continuous-time systems in Section 3, and we shall consider these systems in the last section of the chapter. As for the second assumption, it is possible to take the view that if the law of change in a given dynamical system varies with time, then the system also possesses a "super law" that describes how the given law changes with time. In mathematical terms, this "super law" may be seen as follows:

Suppose that Ω is a set. For every element ω in Ω, let T_ω be a mapping of X into itself. In other words, for each ω, suppose that $T_\omega : X \rightarrow X$. Suppose also that S is a given mapping from Ω to Ω. We shall agree that if T_ω is the law governing the change of states of the given system at time t, then the law governing the system at time $t + 1$ is $T_{S(\omega)}$. With this interpretation we can regard S as the super law that governs changes in the law T.

We shall now show that a dynamical system of this sort can be replaced by a new (but more complicated) system in which the states change according to a law that does not vary with time. The states in the new system will be the ordered pairs (ω, x) that can be made by taking ω from Ω and x from X, and the set of these ordered pairs will be written as $\hat{X}$. Thus a state of the new system determines a state x of the old system together with the law T_ω that governs X at the present moment. In the state space X, we define a new law $\hat{T}$ that combines the variable law T_ω and the super law S. We define $\hat{T} : \hat{X} \rightarrow \hat{X}$ by

$$\hat{T}(\omega, x) = (S(\omega), T_\omega(x)).$$

We see that the orbit of any state (ω, x) in the new system is

$$(\omega, x), (S(\omega), T_\omega(x)), (S^2(\omega), T_{S(\omega)}(T_\omega(x))), \ldots$$

This new system $(\hat{X}, \hat{T})$ is called the *skew product* of the original system and the super law $S : \Omega \rightarrow \Omega$. This mathematical device of forming the skew product comes to terms, at least in principle, with the problem of changing laws of change. It reduces the system with a variable law to a new system with a constant law, and even though this process depends upon our very strong assumption that the original law varies according to a "super law," we may consider this device to be an argument in principle. We may therefore take this device as a justification for our decision to restrict ourselves to systems in which the states change according to a law that does not vary with time.

In Section 1 we shall investigate those dynamical systems that have *finite* state spaces. In this special case, the character of the orbits of the system can be understood completely. The results about "finite periodicity" that have been obtained for finite state spaces then serve as a motivation for the study of more general systems. We shall conclude Section 1 with a glimpse of the theory of automata. In Section 2 we shall display a famous dynamical system with a (countably) infinite state space. This system, known as the *Game of Life*, was conceived by J. H. Conway in Cambridge during the sixties. We shall look at a number of examples of possible states in this system, and these will provide us with some insight as to the abundant wealth of possible kinds of orbits. The insights that we can gain from *Game of Life* also extend beyond mathematics into such sciences as biology.

In Section 3 we present some other dynamical systems that have played an important role in mathematical research during the course of the last hundred years. These include *circle rotations* (also known as Kronecker systems), *baker ("French dough") transformations*, and *Smale's horseshoe*. Then in Section 4 we present a particularly versatile dynamical system known as the *shift*. In looking at all these examples, we shall attempt to gain an idea of the basic results in the mathematical theory of dynamical systems. The dynamical systems in Section 1 to 5 are mostly discrete-time systems, but in Section 6 we shall look at two typical but contrasting themes that concern those dynamical systems in which the time variable varies continuously. These two themes lead to the notions of *stability* and *instability*. The results that we touch upon in this section are of importance for both celestial mechanics and statistical mechanics, and they are related to a variety of famous old problems.

We suggest that the reader read this chapter in parallel with Ekeland [1984]. Students of mathematics may wish to look at some deeper or more detailed work such as the volumes *Selecta Mathematica IV, V* (Springer Verlag 1972, 1979).

§1   *Dynamical Systems with a Finite Number of States*

In this section, we investigate those dynamical systems whose spaces are finite sets. We shall obtain a very simple result:

> If the state space is finite, then every orbit is either periodic or it becomes periodic after a finite number of steps.

Such orbits can be seen in Figure VI.1.1.

1.1   *The Injective Case*

We first settle the case in which the mapping $T : X \to X$ is *injective*—in other words, when T sends different states to different states. More precisely, the mapping T is injective when

$$x \neq y \Rightarrow T(x) \neq T(y).$$

It is worth noticing that if T is injective, so is T^n for every positive integer n. In other words,

Periodic orbits (cycles)   Becoming periodic

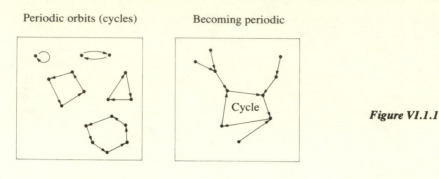

Figure VI.1.1

$$x \neq y \Rightarrow T^n(x) \neq T^n(y)$$

for every positive integer n. This fact may be proved easily by induction, and we leave the proof to the reader.

Theorem 1.1. If X is a finite set and T is an injective mapping from X into X, then every orbit in the system (X,T) is periodic. More precisely, for every state x from X, there is an integer $d \geq 1$ (called the period of x) such that the points $x, T(x), \ldots, T^{d-1}(x)$ are all distinct, but $T^d(x) = x$.

Note that if d is the period of x, then

$$T^{d+1}(x) = T(T^d(x)) = T(x).$$

Continuing in this way, we can see that for every natural number n we have $T^{d+n}(x) = T^n(x)$. In the special case $d = 1$ we have $T(x) = x$, and so $T^n(x) = x$, for every natural number n. In this case we say that x is a *stationary point* of the system (X,T). Alternatively, we say that x is an *invariant point* or a *fixed point* of the system. We can regard *statics* as the study of fixed points of a dynamical system.

Proof of Theorem 1.1. We suppose that X consists of n states and that x is one of these. Consider the first $n + 1$ states in the orbit of x. These are

$$x, T(x), \ldots, T^n(x).$$

We conclude from the pigeonhole principle (Theorem III.6.1) that these $n + 1$ states cannot all be different. Therefore, we can find two nonnegative integers i and j such that $i < j$ and $T^i(x) = T^j(x)$. Since T^i is injective, we conclude that $x = T^{j-i}(x)$. We now define d to be the least positive integer satisfying $T^d(x) = x$. We conclude that the points $x, T(x), \ldots, T^{d-1}(x)$ are all distinct and that the orbit of x has the form

$$x, T(x), \ldots, T^{d-1}(x), x, T(x), \ldots, T^{d-1}(x), \ldots$$

We have therefore shown that every orbit of the system (X,T) is periodic.

Note that all the points in a given orbit have the same period, and that two distinct orbits must always be disjoint. We can therefore say that X is partitioned into its periodic orbits. A periodic orbit of period d may also be called a *cycle* of *length d*.

1.2   The General Case

We suppose now that X is a finite set, and that $T : X \to X$. In this case, the mapping T is not necessarily injective. We define X_1 to be the set of all states that have the form $T(x)$ for some x from X. In set-theoretic notation,

$$X_1 = \{T(x) \mid x \in X\}.$$

Clearly $X \supseteq X_1$. Having defined X_1, we define

$$X_2 = \{T(x) \mid x \in X_1\},$$

and, continuing in this way (inductively), we obtain a sequence of subsets $X_1, X_2, \ldots$ of X such that

$$X \supseteq X_1 \supseteq X_2 \supseteq \ldots \supseteq X_t \supseteq \ldots$$

If X has n states and X_t has n_t states, then we have

$$n \geq n_1 \geq n_2 \geq \ldots$$

We note that every one of the sets X_t must contain at least one state. In other words, $n_t \geq 1$ for every t. It therefore follows that for some time t we have

$$n_t = n_{t+1} = n_{t+2} = \ldots$$

from which we deduce that

$$X_t = X_{t+1} = X_{t+2} = \ldots$$

We now define $Y = X_t$. Since $n_t = n_{t+1}$, no two different states in Y can be sent by T to the same state. In other words, if we ignore the states that lie outside the set Y, we see that the mapping T is injective within Y. We can therefore apply Theorem 1.1 and conclude that Y is partitioned into periodic orbits. Since every state in the set X must arrive in Y after at most t steps, we have proved the statement at the beginning of this section.

Theorem 1.2. If X is a finite set and $T : X \to X$, then there is at least one periodic orbit in X. Furthermore, given any state x in X, the orbit of x enters and remains inside exactly one periodic orbit.

We can express this result simply by saying that if a state space is finite, then every one of its orbits is eventually periodic.

1.3   A Glimpse of the Theory of Automata

If we consider several mappings $T_1, \ldots, T_r$ of a finite set X into itself, then we obtain what is called a (finite) *automaton* in mathematics. We may interpret an automaton as an apparatus whose potential states are represented by the elements in the set X. The apparatus has r buttons. If we press button number k while the apparatus is in state x, then it jumps into the state $T_k(x)$. If we then press button number j, the apparatus changes into the state $T_j(T_k(x))$. If there is only one button, or if the mappings T_k are all the same, then we may define $T = T_1$, and the system reduces to a dynamical system

(X,T) of the type we considered previously. In this case, every orbit is eventually periodic. In general, however, it is possible to press the buttons of an automaton in such a way that some of the orbits are not eventually periodic.

One may perhaps interpret the world as a dynamical system, an automaton with only one button that is pressed every time the clock ticks. This way of thinking about the world is an extreme form of what is called Deism in philosophy. If we allow only a finite number of possible states of the world, then Theorem 1.2 holds for the system and the world is drowned in periodicity. The idea of *eternal return* has, in fact, been suggested over and over again in history, and practically always in combination with assumptions about the finiteness of the world. We shall give special emphasis to the notion of return or recurrence throughout this chapter.

§2   *Game of Life*

Game of Life is a dynamical system that was conceived in the sixties by J. H. Conway in Cambridge. Over the past twenty years, Game of Life has attracted the attention of many scientists, especially biologists; see, for example, Eigen-Winkler [1975]. A comprehensive presentation of Game of Life can be found in Berlekamp-Conway-Guy [1982], Part 4. Both the state space X and the mapping T from X to X that make up Game of Life are a little too complicated to be given here in full mathematical rigor, but we can obtain a fairly good intuitive idea of what the system is by looking at it visually.

2.1   *The State Space of Game of Life*

In order to describe the state space X of Game of Life, we imagine that the plane is decomposed into square cells as shown in Figure VI.2.1. We imagine that each of these cells can be left vacant or can be filled with a black disc. A state in the system is a distribution of black discs into a finite number of the cells. The number of black discs can even be zero. In this case, all the cells are left vacant, and we refer to this state as the *null state*. Looking at the idea of a state in another way, we might say that a state is

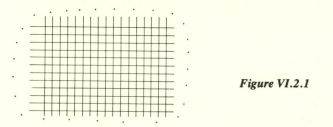

Figure VI.2.1

a list of that finite number of cells that are occupied. Figure VI.2.2 illlustrates some examples of states in Game of Life.

We shall now observe that the set of all possible states in this system is countable. We

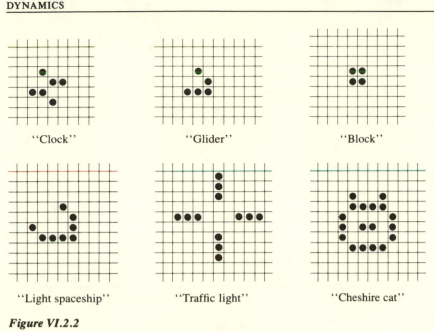

"Clock"   "Glider"   "Block"

"Light spaceship"   "Traffic light"   "Cheshire cat"

Figure VI.2.2

start by noting that only a finite number of states can have all their occupied cells included in any given square. Next we choose a point c that lies on the corner of one of the cells, and for each natural number n we write Q_n for the square with center c and a side of length $2n$ units, where each cell has a side of length one unit. We can now enumerate the set X of all states in Game of Life by listing all the states whose occupied cells all lie in Q_1, then listing all the states whose occupied cells all lie in Q_2 but not all in Q_1, then listing all the states whose occupied cells all lie in Q_3 but not all in Q_2 . . . Continuing in this fashion, we may enumerate all the states in the system.

Now that we know that there are only countably many states in Game of Life, we could change our description of the game in such a way that the state space becomes the set of natural numbers. However, we shall not do this, because a description of this sort would utterly destroy the geometric flavor of Game of Life.

2.2   The Transition Law T of Game of Life

In order to describe the mapping $T : X \to X$ for Game of Life, we shall say how the image state $T(x)$ needs to be constructed for each state x. In other words, we have to describe, for each state x, which of the cells are to be filled with black discs in the state $T(x)$ and which cells are to be left empty. For this purpose, we consider a given cell with its eight neighboring cells as shown in Figure VI.2.3.

In this figure, the letter N stands for north, NE stands for northeast, and so on. The *number of neighbors* that a given cell has in state x means the number of these eight neighboring cells that are occupied by black discs in state x. The state $T(x)$ is now constructed from x according to the rules in the table. These rules describe the state $T(x)$ for any given state x.

NW	N	NE
W	$\otimes$	E
SW	S	SE

Figure VI.2.3

Rule no.	Precise version	Shortcut version
I	A cell that is occupied under state x will be occupied under $T(x)$ if and only if it has 2 or 3 neighbors that are occupied under x	2 or 3 neighbors keep you alive
II	A cell that is vacant under state x will be occupied under state $T(x)$ if and only if it has precisely 3 neighbors that are occupied under x	3 neighbors create life
III	If neither II nor III apply to a given cell, this cell will be vacant under state $T(x)$	You die if you are alone, and you die in the crowd

2.3 The Life Stories (Orbits) of Some Configurations (States)

In this subsection we display some information about the orbits of some particular states of Game of Life. We rely heavily on Berlekamp-Conway-Guy [1982]. We have labeled the states using the names by which they have come to be called in recent work in this field. Occupied cells in state x that become vacant in the state $T(x)$ are marked as an open circle, while empty cells in state x that become occupied in state $T(x)$ are marked as black discs. Much of the information given in this subsection is computer-generated.

- (a) **Configurations That Remain Constant:** These are also called stable, stationary, or fixed points of the mapping T. The simplest of these is, of course, the null state. In the null state, the cells remain empty by rule III. Fig. VI.2.4 illustrates some more stable states.
- (b) **Configurations (States) with Period 2:** Figure VI.2.5 illustrates some of these states.
- (c) **Configurations (States) with Period 3:** Figure VI.2.6 illustrates some of these states.

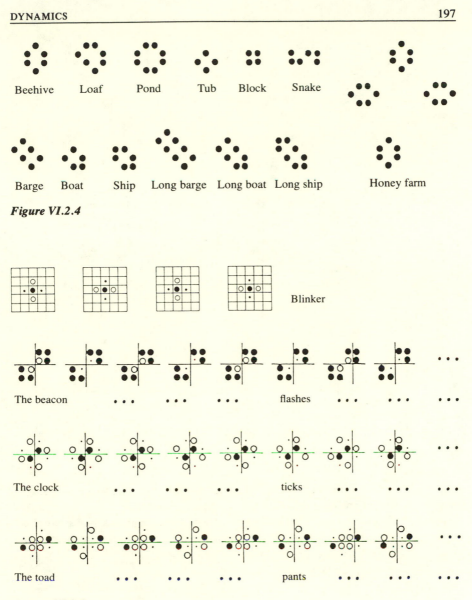

Figure VI.2.4

Figure VI.2.5

(d) **A Configuration with Period 15:** An example of a configuration of this type is shown in Figure VI.2.7.

(e) **Configurations That Die Out:** These are configurations that lead to the null state after a finite number of steps. We illustrate two examples of these "mortal" states in Figure VI.2.8. An interesting question to ask is, for which natural numbers n will the state consisting of n occupied cells in a

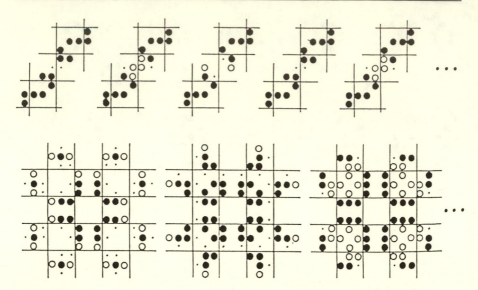

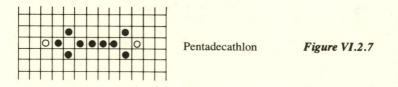

Figure VI.2.6

Pentadecathlon   Figure VI.2.7

row be mortal? See Figure VI.2.9. Some values of n for which this state is mortal are 1,2,6,14,15,18,19.

(f) Configurations That Become Stable or Periodic after a Finite Number of Steps: The state that consists of n occupied cells in a row has the following properties depending upon the value of n:

$n = 3$ gives us the "blinker," with period 2.

 Dies after 2 steps

Dies after 4 steps

Figure VI.2.8   Figure VI.2.9

n = 4   becomes the (stable) "beehive" after 2 steps.

n = 5   becomes the "traffic light" with period 6 steps.

n = 7   becomes the stable honey farm after 14 steps.

n = 8   becomes four blocks plus four beehives: stable.

n = 9   becomes two traffic lights with period 2.

n = 10 becomes the pentadecathlon with period 15.

(g) **The Glider** reproduces its shape after four steps but moves slowly to the southeast. See Figure VI.2.10.

Figure VI.2.10

(h) **The Glider Gun** was found by an MIT team led by R. W. Gosper in November 1970 with the aid of a computer. The glider gun regains its shape every 30 steps, but produces one glider during that period. Thus the whole configuration becomes larger and larger in the course of time. Figure VI.2.11 is taken from the Springer Mathematics Calendar of 1977, and shows some of the 30 steps.

Theorem 2.1. Game of Life can simulate any computer. For a sketch of the proof of this result, see Berlekamp-Conway-Guy [1982], Part 4.

2.4   The Garden of Eden Theorem

A configuration x in Game of Life is called a *Garden of Eden configuration* or an *orphan* if it has no predecessor—in other words, if there is no state y such that $x = T(y)$. The Garden of Eden theorem states simply that there is at least one Garden of Eden in Game of Life. As a matter of fact, there are many Gardens of Eden in Game of Life.

Proof. For some very large natural number n that we shall choose later, we consider a square in the plane that is composed of n^2 smaller squares of size 5×5. We shall call these squares of size 5×5, the *components* of the large square. We shall call the large square the "$5n \times 5n$ square," and we may visualize it as it appears in Figure VI.2.12.

If we cut off a border "seam" of width 1 around this $5n \times 5n$ square, we are left with what we shall call the $(5n - 2) \times (5n - 2)$ square. There are

$$2^{(5n-2)^2}$$

possible ways of distributing black discs into the $(5n-2)^2$ cells in the $(5n-2) \times (5n-2)$ square while leaving all other cells in the plane empty. In looking for predecessors of these configurations, we need consider only those that leave empty all cells outside the

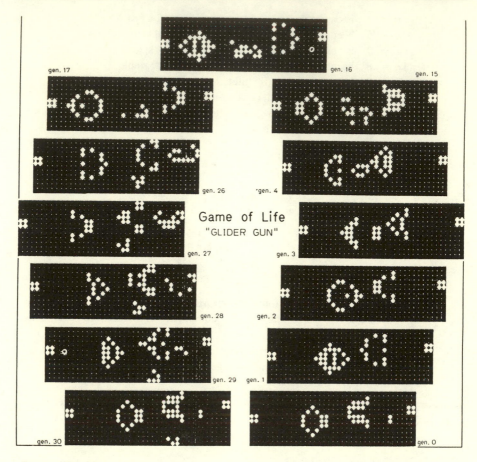

Figure VI.2.11

$5n \times 5n$ square but do not leave empty any of the 5×5 components of the $5n \times 5n$ square. In fact, if any of the 5×5 components is empty, then we may insert a black disc into its central cell and thus obtain a configuration that has the same successive state as the one with which we started. See Figure VI.2.13. We therefore need consider only $(2^{25} - 1)^{n^2}$ possible candidates as predecessors for the

$$2^{(5n-2)^2} = 2^{25n^2 - 2n + 4}$$

possible configurations in the $(5n-2) \times (5n-2)$ square.

We shall show that if n is sufficiently large, then there are not enough configurations to provide at least one predecessor for each of the configurations in the $(5n-2) \times (5n-2)$ square. To make this observation, we begin by noticing that the number $2^{25} - 1$ can be written in the form 2^a, where a is a number that is a little less than 25. In fact,

$$a = 24.999999957004337 \ldots$$

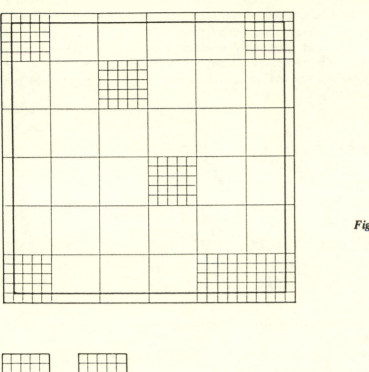

Figure VI.2.12

 as for successor   *Figure VI.2.13*

Now since

$$\frac{25n^2 - 2n + 4}{an^2} = \frac{25 - 2/n + 4/n^2}{a},$$

and since the number $25 - 2/n + 4/n^2$ can be made as close as we like to 25 by making n large enough, it follows that for all sufficiently large values of n, we have

$$25n^2 - 2n + 4 > an^2.$$

(Using a computer, it may be shown that the latter inequality holds for $n \geq 465163200$). We have therefore shown that if n is sufficiently large, then

$$2^{25n^2 - 2n + 3} > (2^{25} - 1)^{n^2},$$

which is what we needed to show to guarantee that the $(5n - 2) \times (5n - 2)$ square contains some orphans.

The idea behind the preceding proof may also be used to prove "Garden of Eden" theorems for a large class of other dynamical systems called *cellular automata*. These are dynamical systems whose states are defined by inserting symbols into cells, where the symbols are taken from a finite set ("finite alphabet") of possibilities. The transition law T in such systems is defined by cell occupancy changes that depend on the situation in certain neighboring cells only (see Arbib [1966], [1969]). As we may deduce from our previous considerations, every square of side length 2325816000 in Game of Life must contain a Garden of Eden, but there may also be smaller squares that do the job. Figure VI.2.14 shows a fine little orphan in Game of Life. This example has been checked using a computer.

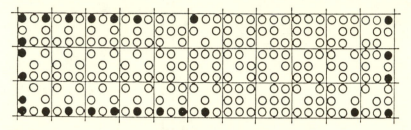

Figure VI.2.14

§3 *Some Further Dynamical Systems*

In this section we present three more dynamical systems that have played an important role in mathematical research over the past few decades.

3.1 *Circle Rotation (Kronecker [1884])*

In this examples we shall take the set X to be a circle of radius 1, and we shall take T to be a rigid rotation of the circle through a given angle. See Fig. VI.3.1. The orbit of an arbitrary point x on the circle is obtained by rotating the point x through the given angle over and over again. See Figure VI.3.2. There are two separate cases that need to be considered in our discussion of this dynamical system.

Case I, Periodic Case: This case arises if and only if a point x on the circle will return to its starting point after a certain number of rotations T. If t is the smallest number of rotations that will return a given point x to its starting point, then it is clear that the point x will also return to its starting point after nt rotations, where n is any positive integer. Furthermore, since the rotation of the circle is rigid, every point on the circle will return to its starting point after the same number of rotations.

Case II, Kronecker's Case: When Case I does not apply, no point on the circle can ever return to its starting point. In this case it is clear that no point on the circle can be sent to the same place at two different times, and thus the orbit of each point is an infinite set. We may therefore argue that if x is a point on the circle, then there must be two points on the orbit of x that lie within a distance of 1/1000 of each other. (Note that if any set of points on the circle contains more than 2000π members, then there must be two members of the set that lie within a distance 1/1000 of each other.) We conclude that if x is

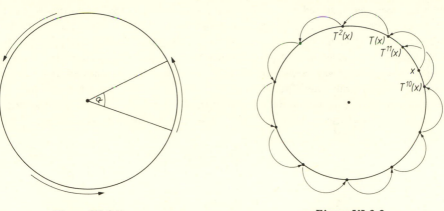

Figure VI.3.1   *Figure VI.3.2*

any point on the circle, then x is sent in some "time" $t > 0$ to a point y that lies within a distance 1/1000 of x. By setting our clock to the times t, $2t$, $3t$, . . . , we can move around the circle with step width $\leq 1/1000$ and obtain a set of points on the orbit of x such that every point on the circle lies within a distance 1/1000 of one of these. By replacing the number 1/1000 by an arbitrary positive number ϵ, the same argument allows us to deduce the following theorem of Kronecker:

Theorem 3.1. (Kronecker [1884]). If, under a given rigid rotation of the circle, no point ever returns to its starting point, then every orbit contains points that lie arbitrarily close to any given point. In short, every orbit is *dense*.

This result of Leopold Kronecker (1823–1891) is sometimes known as Kronecker's *density theorem*. It can easily be carried over to other situations. For example, the dynamical system known as the *Kronecker flow* in the unit square is obtained from the circle rotation by expanding it into a continuous movement as shown in Figure VI.3.3. The cylinder is

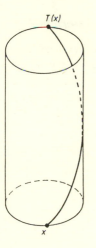

Figure VI.3.3

then cut open and flattened out, as shown in Figure VI.3.4. Each point undergoes a uniform straight-line motion until it reaches either the top of the square or its right side. When a point reaches the right side of the square, it disappears there and reappears at the opposite point on the left side. A point that reaches the top of the square disappears and reappears in a similar way. See Figure VI.3.5. When this system is obtained from a circle rotation that falls into Case I, the orbit of each point is a lattice of a finite number of segments in the unit square that are equidistant from one another (''zebra''). If the circle rotation falls into Case II, then each orbit in the square covers a dense subset of the square.

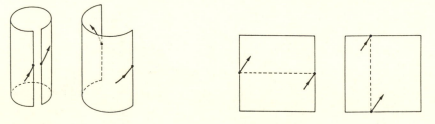

Figure VI.3.4   *Figure VI.3.5*

Before we leave this subsection we mention that in 1916, Hermann Weyl (1885–1955) proved a result about the circle rotations that is much stronger than Kronecker's density theorem. Weyl showed that if the circle rotation falls into Case II, then in the long run, every point must visit any given arc in the circle with a frequency that is proportional to the length of the arc: the ''zebra'' turns into a uniformly grey ''donkey.'' We shall not give a precise statement of this result of Weyl, which is known as *Weyl's equidistribution theorem* (Weyl [1916]) and which has given birth to an entire mathematical theory known as *equidistribution theory*. An excellent introduction to this theory, for mathematicians, may be found in Hlawka [1979].

3.2   The French Dough Transformation (The Baker's Transformation)

The word ''transformation'' in the name of this subsection has the same meaning as ''function'' or ''mapping.'' Like the variation of the circle rotation that we discussed in the preceding subsection, the baker's transformation acts in the unit square X with its top and right edges removed. Using the language of analytic geometry we can describe the set X as the set of all ordered pairs (x, y) of real numbers such that $0 \leq x <$

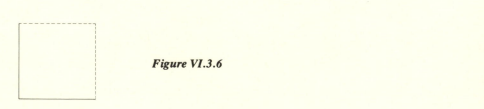

Figure VI.3.6

1 and $0 \leq y < 1$. Note that by requiring these numbers to be strictly less than 1, we have automatically removed the top and right edges of the square. The mapping T in this example acts in X in much the same way that a baker kneads his (or her) dough:

    (a) Halve it vertically.
    (b) Flatten the halves.
    (c) Put the right half on top of the left half.

See Figure VI.3.7.

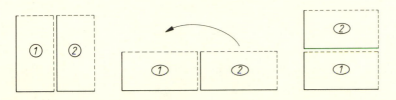

Figure VI.3.7

We may express this transformation analytically as follows:

    (a) left half: $0 \leq x < \dfrac{1}{2}, \; 0 \leq y < 1$

        right half: $\dfrac{1}{2} \leq x < 1, \; 0 \leq y < 1$

    (b) flatten $x \to 2x, \; y \to \dfrac{1}{2}y$

        left half: $0 \leq 2x < 1, \; 0 \leq \dfrac{1}{2}y < \dfrac{1}{2}$

        right half: $1 \leq 2x < 2, \; 0 \leq \dfrac{1}{2}y < \dfrac{1}{2}$

    (c) put right half on top of left half:

$$2x \to 2x - 1, \; \frac{1}{2}y \to \frac{1}{2}y + \frac{1}{2}$$

        hence: $0 \leq 2x - 1 < 1,$

$$\frac{1}{2} \leq \frac{1}{2}y + \frac{1}{2} < 1.$$

This mapping T can be expressed in the form

$$T(x,y) = \begin{cases} \left(2x, \dfrac{1}{2}y\right) & \text{whenever } 0 \leq x < \dfrac{1}{2} \\[2ex] \left(2x - 1, \dfrac{1}{2}y + \dfrac{1}{2}\right) & \text{whenever } \dfrac{1}{2} \leq x < 1 \end{cases}$$

It is easy to see that the baker's transformation T changes the shape, but not the area, of parts of X. In each action, widths are doubled and heights are halved. Mathematicians describe this property by saying that T *preserves measure*.

The name "baker's transformation" becomes even more plausible if we follow the change of shape of the *upper* and *lower* halves of the square (instead of the left and right halves) under repeated application of the transformation T. See Figure VI.3.8. Samurai swords and Damascene blades are usually made this way.

In §4, we shall connect the baker's transformation with the so-called shift.

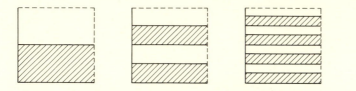

Figure VI.3.8

3.3   Stephen Smale's Horseshoe

This system was published by Smale (b. 1930) in 1965 (Smale [1965]). It consists of an inner deformation T of a disc X such that a given rectangle R is first flattened and then folded onto itself. See Figure VI.3.9. We are not concerned with what happens to other parts of X under the transformation T. What we are interested in is the effect of T on the shaded sections of the rectangle R in Figure VI.3.10. We are also interested in knowing which points of R go into the shaded parts of R. See Figure VI.3.11.

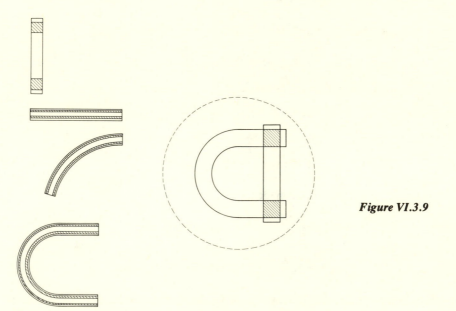

Figure VI.3.9

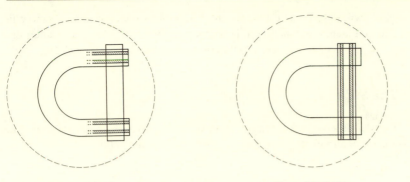

Figure VI.3.10   **Figure VI.3.11**

If we look at the intersections of the two shaded regions, we obtain the region shown in Figure VI.3.12. If we repeat the whole procedure, then we obtain the region shown in Figure VI.3.13.

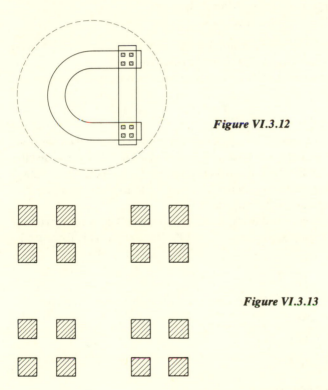

Figure VI.3.12

Figure VI.3.13

Continuing indefinitely in this fashion, we arrive at an intricate set of points. The method we have described is a variation of Cantor's procedure for constructing his discontinuum, and it can be shown, in fact, that the set constructed in this subsection is homeomorphic to Cantor's discontinuum. In topological language we say that this set *is* a "Cantor discontinuum." In §4 we shall connect Smale's horseshoe with the so-called shift.

§4   *The Shift*

The *shift* is a dynamical system of universal importance. In this case the state space X is the set of all infinite 0-1-sequences. We shall use the notion

$$x = x_0 x_1 x_2 \ldots$$

(written without commas) for a 0-1-sequence, where $x_n = 0$ or 1 for every nonnegative integer n. We discussed 0-1-sequences in Section 7 of Chapter III, and we saw some interesting examples like the ones shown in the following table:

$$0\ 0\ 0\ 0\ 0\ 0\ 0\ 0\ 0 \ldots$$
$$0\ 1\ 0\ 1\ 0\ 1\ 0\ 1\ 0 \ldots$$
$$1\ 0\ 1\ 0\ 1\ 0\ 1\ 0\ 1 \ldots$$
$$0\ 1\ 1\ 0\ 1\ 0\ 0\ 1\ 1 \ldots \text{(Thue-Morse sequence)}$$
$$0\ 1\ 0\ 0\ 1\ 0\ 0\ 0\ 1 \ldots \text{(rarefied ones)}$$

We also proved in Chapter II that the set X of all 0-1-sequences is uncountable. As you may recall, we proved this result using the Cantor diagonal method.

In the set X of all 0-1-sequences, the *shift* mapping is defined quite naturally as follows:

$$T(x_0 x_1 \ldots) = x_1 x_2 \ldots$$

The shift mapping throws away the first member x_0 of the sequence and then shifts all the remaining entries one step to the left. The set X is also called the *shift space*, and the name "shift" is also used for the whole dynamical system (X,T). It is possible to replace the symbols 0 and 1 by any other two different symbols, or to use more than just two symbols. However, we shall stick to 0-1-sequences here.

The shift (X,T) is universal in the following sense. Suppose that (Y,S) is any dynamical system. Split Y into two mutually disjoint parts Y_0 and Y_1. If y is any state in the space Y, we consider its orbit

$$y, S(y), S^2(y), \ldots$$

Every one of the states $S^t(y)$ in this orbit is either in Y_0 or in Y_1, and we can define a symbol x_t by

$$x_t = \begin{cases} 0 \text{ if } S^t(y) \text{ is in } Y_0 \\ 1 \text{ if } S^t(y) \text{ is in } Y_1 \end{cases}$$

This definition provides us with a sequence

$$x = x_0 x_1 x_2 \ldots$$

that belongs to the space X. We shall call this sequence the 0-1-sequence *representing y* (for the given partition of Y into Y_0 and Y_1). In the example illustrated in Figure VI.4.1, S is a rigid rotation of a circle through an angle α (see Subsection 3.1), and the point y has a representing sequence

$$x = 0\ 0\ 0\ 1\ 1\ 1\ 0\ 0\ 0\ 0\ 0\ 0\ 0\ 1\ .\ .\ .$$

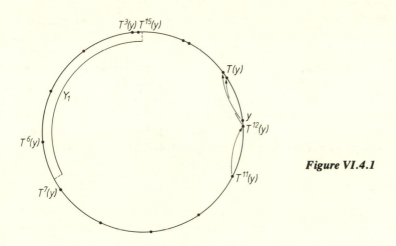

Figure VI.4.1

In general, if a point y in a state space Y is represented by a sequence

$$x = x_0\, x_1\, x_2\, .\, .\, .\, ,$$

then the point $S(y)$, whose orbit is

$$S(y),\ S^2(y),\ S^3(y),\ .\, .\, .\, ,$$

is represented by the sequence

$$T(x) = x_1\, x_2\, .\, .\, .$$

We see, therefore, that if we represent the states in the space Y (with respect to a given partition of Y into subsets Y_0 and Y_1), then this process also represents the mapping S in the system (Y,S) by the shift mapping T.

Thus we see how the shift can represent all dynamical systems. In this sense, the shift is a universal dynamical system. We should not, however, overlook the fact that the representation of a given dynamical system (Y,S) depends heavily on the particular partition Y_0, Y_1 that we choose for the space Y. The representation of the system (Y,S) according to some partitions will be much more fruitful than it is with others. For an extreme case of a partition we could choose $Y_0 = Y$ and choose Y_1 to be the empty set. According to this partition, every state y in the space Y will be represented by the same sequence

$$0\ 0\ 0\ 0\ 0\ 0\ 0\ .\, .\, .\, ,$$

and the representation will provide us with no information at all about the system (Y,S). We might call a given representation of the system (Y,S) *faithful* if the partition Y_0, Y_1 has been chosen in such a fashion that any two different states y and y' in Y have different representing sequences in X. More precisely, a given representation is faithful if for any two different states y and y' in Y, there is at least one integer t such that one of the states $S^t(y)$ and $S^t(y')$ lies in the set Y_0 and the other lies in Y_1. When this happens, if x and x' are the representing sequences of y and y', respectively, then one of the numbers x_t and x'_t is 0 and the other is 1. In other words, the sequence x and x' differ in the t th position.

If Y_0 and Y_1 have been chosen in this way, then we say that Y_0 and Y_1 *separate the orbits* of the given system (Y,S). Alternatively, we say that the partition of S into S_0 and S_1 is a *generator* for the system. Not every dynamical system has a generator. There is an intricate theory that provides necessary and sufficient conditions for a given dynamical system to have a generator, but we cannot present the details of this theory here. We mention only that the notion of *entropy* plays an important role in this theory. Entropy measures the degree of *chaos* in the dynamical system (Y,S). If we want to represent a dynamical system (Y,S) with sequences that involve symbols from a finite "alphabet" like

$$0,1,2, \ldots , n,$$

then a faithful representation is possible as long as the system (Y,S) has "finite entropy." Systems with infinite entropy require the use of infinite alphabets. The ideas at which we have hinted here can be translated into rigorous mathematics, but they require the use of specialized techniques such as the branch of mathematics known as *measure theory*.

In the case of the circle rotations, Case II (see Subsection 3.1), the entropy of the system is zero and the alphabet $\{0,1\}$ is sufficient. In this case one may use Kronecker's theorem to show that if the circle is split into any two arcs Y_0 and Y_1, the representation will be faithful. Indeed, if y and y' are two different points on the circle line, iterated circle rotation (Case II) will bring the shorter of the two arcs with endpoints y, y' into practically every position on the circle, hence also into a position where $T^t y$ is in Y_0 and $T^t y'$ is not in Y_0.

In addition to the shift system (X,T) that we have already discussed in which the sequences have the form

$$x = x_0 x_1 x_2 \ldots ,$$

it is also possible to define a shift system using doubly infinite sequences of the form

$$x = \ldots x_{-2} x_{-1} x_0 x_1 x_2 \ldots$$

When we use sequences of this sort, then the shift mapping T that moves each character one place to the left is a one-one mapping because it does not throw away any symbol from a sequence in order to make the shifted sequence. The dynamical system (X,T) thus obtained is called the *bilateral shift*. This system is related to the baker's transformation (Subsection 3.2) in a very interesting fashion. The relationship is established by using what are known as *dyadic expansions* of real numbers that lie in the interval $[0,1]$. We shall now take a moment to explain briefly what is meant by the dyadic expansion of a given number in the interval $[0,1]$.

Suppose that $u \in [0,1]$. The dyadic expansion of u is the 0-1-sequence

$$x_0 \, x_1 \, x_2 \, x_3 \, \ldots$$

in which the characters x_i have the following properties:

$$x_0 = 0 \text{ if } 0 \le u < \frac{1}{2}$$

$$x_0 = 1 \text{ if } \frac{1}{2} \le x \le 1.$$

Once we have determined in which of these two intervals of length $1/2$ the point u lies, we take $x_1 = 0$ if u lies in the left half of *this* interval and take $x_1 = 1$ if u lies in the right half. We continue this process. It is easy to see that any two different numbers in the interval $[0,1]$ must have different dyadic sequences. It can also be shown that every 0-1-sequence is the dyadic sequence of some number u in the interval $[0,1]$.

Now to see the relationship between the bilateral shift and the baker's transformation, we shall look at a typical point (u,v) in the unit square. In other words, (u,v) is an ordered pair of numbers u and v that lie in the interval $[0,1]$. We shall write the dyadic expansions of u and v in the form

$$x_0 \, x_1 \, x_2 \ldots \qquad \text{and} \qquad x_{-1} \, x_{-2} \, x_{-3} \ldots ,$$

and having done so, we shall paste these two sequences together to form the single bilateral sequence

$$x = \ldots x_{-2} \, x_{-1} \, x_0 \, x_1 \, x_2 \ldots$$

We shall now investigate the effect of the baker's transformation T on this bilateral 0-1-sequence. The question that we have to ask is, what sequence would be obtained from the point $T(u,v)$ in the unit square if a given sequence x is obtained from the point (u,v)? Now in the event that the point (u,v) lies in the left half of the square, the baker's transformation sends (u,v) to the point $(2u, v/2)$. It may easily be verified that the dyadic expansions of $2u$ and $v/2$ are

$$x_1 \, x_2 \ldots \qquad \text{and} \qquad 0 \, x_{-1} x_{-2} \ldots = x_0 \, x_{-1} \, x_{-2} \ldots ,$$

and it follows that the bilateral sequence that corresponds to the point $T(u,v)$ is the same as the sequence corresponding to (u,v) but shifted one place to the left. A similar argument may be used to show that the same result holds if the given point (u,v) lies in the right half of the unit square. We leave the details of this proof as an exercise.

The shift spaces (X,T) and (X,T) can be used as models for perfectly random repetitions of a simple binary experiment like tossing a coin (0 for heads and 1 for tails). Therefore, the relationship that we have described between the baker's transformation and the bilateral shift can be interpreted with the following statement:

> The baker's transformation produces chaos in the unit square as if it were a purely random process.

A considerable amount of mathematical technique is required in order to transform this statement into precise mathematics. If we replace the unit square by the figure shown in

Figure VI.4.2, obtained by removing middle thirds from an interval in the way the Cantor discontinuum is constructed (Cantor [1883]), then we arrive at a situation that is similar to the one obtained in Subsection 3.3, where we discussed Smale's horseshoe. Here, too, we may work with dyadic expansions. All we need to do is replace the words "left half"

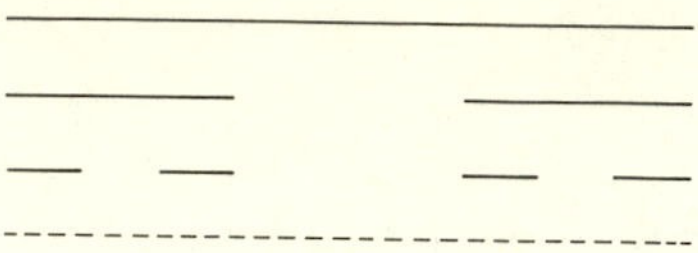

Figure VI.4.2

and "right half" by "left third" and "right third." The relationship between Smale's horseshoe and the bilateral shift suggests the following statement:

> Smale's horseshoe produces chaos in certain parts of its domain, as if it were a purely random process.

Moser [1973] has discovered a large number of horseshoe-like phenomena that exist in certain kinds of mechanical systems, thus explaining their random-like behavior. See also Jacobs [1978a].

The rest of this section is devoted to a discussion of recurrence phenomena in shift space.

4.1   Fixed Points

In the shift space X there are precisely two fixed points of the mapping T, namely

$$0\ 0\ 0\ 0\ \ldots$$

and

$$1\ 1\ 1\ 1\ \ldots$$

These are the only points at which the dynamical system behaves like the most primitive kind of system: a *static* system.

4.2   Points of Period Two

There are precisely two 0-1-sequences in X that have period 2, namely

$$x = 0\ 1\ 0\ 1\ \ldots$$

and

$$y = 1\ 0\ 1\ 0\ \ldots$$

Since $T(x) = y$ and $T(y) = x$, the mapping T oscillates between x and y.

4.3   Longer Periods

Given any natural number n, we can construct a sequence x with period n by taking the symbol 1 in positions $n, 2n, 3n, \ldots$ and the symbol 0 in all the other places.

$$\underbrace{0\,0\ldots0\,1}_{n-1} \quad \underbrace{0\,0\ldots0\,1}_{n-1} \quad \underbrace{0\ldots}_{}$$

$$\underbrace{1\,0\ldots0\,0}_{n-1} \quad \underbrace{1\,0\ldots0\,0}_{n-1} \quad \underbrace{1\ldots}_{}$$

We observe that for a sequence x of this type, the sequences $x, T(x), \ldots, T^{n-1}(x)$ are all different, and $T^n(x) = x$. However, if $n \geq 3$, then there are some other ways of constructing a sequence that has period n. This question opens a playground in which we invite the reader to enjoy being creative.

4.4   Almost Periodicity

In this subsection, we shall become acquainted with a mathematical notion that plays an important role not only in mathematics, but also in celestial mechanics. This notion is a generalization of the notion of periodicity and is called *almost periodicity*.

In order to describe the notion of almost periodicity, we make use of the following two definitions:

(a) Two 0-1-sequences,

$$x = x_0\,x_1\,x_2\ldots$$

and

$$y = y_0\,y_1\,y_2\ldots,$$

are said to be $(1/n)$-*neighbors* if they agree in their first $n+1$ places—in other words, when $x_i = y_i$ for $i = 0, 1, 2, \ldots, n$.

(b) An increasing sequence $t_0 < t_1 < t_2 < \ldots$ of nonnegative integers is said to have *bounded gaps* if there is a number B such that none of the differences $t_1 - t_0, t_2 - t_1, \ldots$ can exceed B. In other words, we require that

$$t_i - t_{i-1} \leq B$$

for every natural number i.

Roughly speaking, a 0-1-sequence

$$x = x_0\,x_1\,x_2\ldots$$

is almost periodic if for every natural number n, the sequence x becomes its own $(1/n)$-neighbor when shifted suitably, with bounded gaps. The precise definition follows:

Definition 4.1. A given sequence x is said to be *almost periodic* if for every natural number n, there is an increasing sequence $t_0 < t_1 < t_2 < \ldots$ with bounded gaps such that for every natural number k, the sequence

$$T^{t_k}(x)$$

is a $(1/n)$-neighbor of x.

In the event that a sequence x is periodic with period d, then by choosing $t_k = kd$ for every natural number k, we obtain

$$T^{t_k}(x) = x,$$

and since every sequence is certainly a $(1/n)$-neighbor of itself, we deduce that x is almost periodic. We may therefore make the obvious-sounding statement that every periodic sequence is almost periodic. As you may have guessed, there are a great many sequences that are almost periodic but not periodic. We shall see some of these in a moment. In sequences of this type, it may be necessary to allow larger and larger gaps in the sequence (t_k) in order to make the sequence

$$T^{t_k}(x)$$

a $(1/n)$-neighbor of x for larger and larger values of n.

We shall now explore the ways in which a sequence x may be almost periodic without being periodic. We shall look at some explicit examples of such sequences and also at an existence theorem. Before we do this, we need to ask ourselves what almost periodicity really means.

In order for a sequence

$$x = x_0 \, x_1 \, x_2 \ldots$$

to be a $(1/n)$-neighbor of itself at time t, the initial block of $n + 1$ terms of the sequence

$$T^t(x) = x_t \, x_{t+1} \ldots$$

has to coincide with the initial block of $n + 1$ terms of x:

$$x_t \, x_{t+1} \ldots x_{t+n} = x_0 \, x_1 \ldots x_n,$$

i.e.,

$$x_t = x_0, \, x_{t+1} = x_1, \ldots, x_{t+n} = x_n.$$

Thus a sequence x is almost periodic if and only if every initial block of x reappears in x with bounded gaps. See Figure VI.4.3.

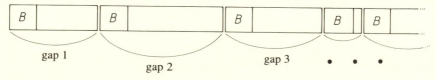

Figure VI.4.3

Equivalently we can say that a sequence x is almost periodic if and only if every finite block that appears anywhere in the sequence must reappear with bounded gaps. The equivalence of these two conditions becomes clear when we realize that any given finite block B that appears in the sequence at all must appear in some initial segment of sufficiently large length. As this initial segment ($=$ block) reappears, so does the given block B. Once again we should bear in mind that the gaps between the reappearances might become larger and larger as the blocks themselves are chosen with larger and larger lengths. The insight that we have gained from these observations should make it easy to verify the almost periodicity of some particular examples of sequences. As a matter of fact, we can take advantage of the results that were obtained in Chapter III (Section 7) and prove the following statement:

The Thue-Morse sequence 01101001 . . . is almost periodic.

We leave as an exercise the task of showing that this sequence is not periodic (see Jacobs [1969a, 1983a]). One may show also that the "Mephisto Waltz"

001001110001001110110110001 . . .

is almost periodic but not periodic; and so is the sequence that is constructed by the following rules:

(1) Put a 0 into every second place.
(2) Put a 1 into every place among the places left over from step 1.
(3) Put a 0 into every place among the places left over from step 2.

   . . .

The sequence obtained this way has the form

010001010100010 . . .

These two kinds of construction can be varied infinitely, leading to a large variety of nonperiodic almost periodic sequences (see Toeplitz [1928], Jacobs [1969a], [1983a], Jacobs-Keane [1969]).

We now state a general theorem concerning the existence of almost periodic 0-1-sequences and sketch its proof. Roughly speaking, the theorem says that every 0-1-sequence approximates some almost periodic sequence arbitrarily closely.

Theorem 4.2. Given any 0-1-sequence

$$x = x_0 x_1 x_2 . . . ,$$

it is possible to find an *almost periodic* sequence $y = y_1 y_2 . . .$ such that if n is any natural number, then there is at least one natural number t such that $T^t(x)$ is a $(1/n)$-neighbor of y.

Sketch of Proof. The theorem states that it is possible to find an almost periodic sequence y such that if n is any natural number, then the block $y_0 y_1 . . . y_n$ of y appears in x as $x_t x_{t+1} . . . x_{t+n}$ for some t. In order to find an almost periodic sequence y that has this property we have to look inside the sequence x for the blocks that will be the initial blocks of the sequence y. The idea is to eliminate everything from x that does not reappear with bounded gaps. Roughly speaking, this can be done as follows:

We shall call a block in x a *critical block* if it reappears either a finite number of times, or else infinitely often, but then only with unbounded gaps. Among the critical blocks in x we select one that has shortest possible length and construct, making skillful use of longer and longer gaps, a "daughter" of x in which this block does not appear at all. "Daughter" means that all initial blocks in this new sequence are blocks from x. Repeating this procedure, we arrive at a "daughter" sequence of x from which all the shortest critical blocks have been removed. Repeating the procedure over and over again, we end up with a "daughter" sequence y of x from which *all* critical blocks of x have been removed (not just the shortest ones). Therefore y is almost periodic. Because of the way in which y has been constructed, it may be seen that x comes arbitrarily close to y at certain times. Details of this proof may be found in Jacobs [1972].

For certain sequences x, it may happen that an "almost periodic daughter" y is visible right away; in such cases we can avoid the clumsy machinery of the preceding argument. For example, the increasing "0-deserts" in the sequence

$$x = 0\ 1\ 0\ 0\ 1\ 0\ 0\ 0\ 1\ \ldots$$

immediately yield $y = 000 \ldots$

Finally, we mention that our use of the term $(1/n)$-neighbor in this subsection was not necessary. We could, instead, have phrased our results in terms of appearances of blocks in the given sequences. However, the term $(1/n)$-neighbor hints at some important generalizations of the theory in which the term "neighborhood" refers to a topological space (see Chapter V). In such generalizations, neighborhood has a geometric flavor. In making such generalizations we enter the mathematical discipline known as *topological dynamics*. We may, for example, repeat the investigations that we have made in this subsection for circle rotations (see Subsection 3.1). In the analogue of this theory for circle rotations, all points on the circle turn out to be almost periodic, even in Kronecker's Case II, in which the points are not periodic. We leave the proof of this assertion as an exercise. In this subsection we have provided a glimpse of the theory of almost periodicity, which arose in the early twentieth century in the study of almost periodicity in celestial mechanics (planetary motion).

§5   *General Results in Dynamics*

In Sections 1 to 4 we became acquainted with some examples of dynamical systems. Now in this section we shall gain a bird's-eye view of some of the basic notions and results in the field of dynamics.

The moment we go beyond the very narrow domain of dynamical systems that have finite state space (which we discussed in Section 1), we need to add some extra properties to the definition in order to produce a fruitful theory. There are several ways in which these extra properties are usually chosen, and each leads to a separate subdiscipline of the field of dynamics.

In the subdiscipline known as *topological dynamics*, the added properties make use of concepts that give a precise meaning to an idea of *closeness* in the state space X. In such spaces X we can speak of a point x as being close to a point y. An example of this type of structure was the space of 0-1-sequences that we considered in Section 4, where the notion of closeness was expressed by the concept of a $(1/n)$-neighbor. In general, the sort of state space in which a notion of closeness is meaningful would be a topological space (see Chapter V). Under fairly general conditions, we can obtain an analogue of Theorem 4.2. All we need is that the topological space X is compact and the transformation T is a continuous function from X to X. Theorem 4.2 can be stated as saying that every point comes arbitrarily close to an almost periodic point. The theory of topological dynamics is a very rich field, but we shall not go into any further details of it here.

Another way of adding special properties to the state space X is to imagine that X is a space that contains a cloud of dust and that in any subset E of X, we can talk about how much of the dust is contained in the set E. A precise formulation of this concept leads to the mathematical notion of a *measure space*, and the corresponding subdiscipline of dynamics in which the state spaces are measure spaces is known as *ergodic theory*. The standard assumption that we make in this subdiscipline is that if E is any "reasonable" subset of X, and if $T(E)$ is the set of all points $T(x)$ for which x is in E, then the amount of dust in $T(E)$ is the same as the amount of dust in E. When this condition holds we say that the way the dust has been distributed in X is *invariant* under T, and we say also that T is a *measure-preserving* transformation on the measure space X. An example of a measure space is the circle in which the "dust" is distributed in such a way that the amount of dust in any arc of the circle is the length of that arc. If we use this measure in the circle, then rotations of the circle are measure-preserving transformations; the baker's transformation (Subsection 3.2) is another important example.

When we have a measure space, the subsets of the space with zero measure have a special role to play. These are the sets that, from the point of view of the given measure, are "essentially empty," and they are known as *null sets*. A condition that holds at all of the points x in a measure space X, except for those in a null set, is said to hold at *almost every* point x in the space, or *almost surely* in X. In many important measure spaces, the finite and countable sets are null sets, but there are usually other more complicated null sets that are uncountable. Using these notions, we can now formulate some of the basic theorems of ergodic theory.

(1) Poincaré's [1890] Recurrence Theorem. If the invariant measure in a given space X has a finite total value (in other words, if there is a finite amount of dust in the whole space), then the following statement holds for every "reasonable" subset E of X:

Almost every point x of E returns to E infinitely often.

These sets that are "reasonable" in the sense of this theorem are the sets that are called "measurable" in the field of measure theory. Note that here, as in Section 1, we are deducing a recurrence statement from a condition of finiteness.

Quite often, in ergodic theory, we combine the given measure-theoretic conditions with topological conditions. If we do so in the preceding theorem, then we obtain the following result:

If the total measure of the space X is finite, then almost every point x in X returns arbitrarily closely to its original position infinitely often.

This statement was transformed by Ernst Zermelo (1871–1953) into an objection against the *entropy-increase theorem (H-theorem)* of Ludwig Boltzmann (1844–1906). To the best of my knowledge, Zermelo's objection has not been fully refuted up to the present day. It is dealt with in Boltzmann [1896] and Ehrenfest-Ehrenfest [1907]. For recent contributions to this topic ("the arrow of time"), see Goldstein [1981] or Zeh [1984].

(2) Birkhoff's [1931] Ergodic Theorem. If E is a measurable subset of a measure space X, then for almost every point x in X, the average amount of time that x spends in E over a time period t approaches a limit as t approaches infinity. The average amount of time that x spends in E over a time period t means the fraction of those numbers s from $\{0,1, \ldots , t-1\}$ for which $T^s(x)$ is in E.

These results, and others of the same type, are the starting point of a world of notions and theorems concerning mixing properties and chaotic behavior of measure-theoretical dynamical systems, among which the results presented in Sections 2–4 play a prominent role (see, for example, Petersen [1983]). Because of the historical origins of ergodic theory, which started with Boltzmann around 1870, some of these results are related to statistical mechanics. For the famous *ergodic hypothesis* of statistical mechanics, there is no rigorous mathematical proof up to this day.

Finally, we may utilize notions and methods of analysis (differential and integral calculus) in order to formulate and prove results about dynamical systems. This subdiscipline of dynamics is sometimes called *differentiable dynamics*, or *dynamics of diffeomorphisms*. Smale's horseshoe (see Subsection 3.3) belongs to this domain, which has become very well developed in recent times.

§6   *Stability and Instability*

We conclude this chapter with a survey of two closely related topics in the theory of time-continuous dynamical systems: *stability* and *instability*. Stability problems have arisen both in celestial mechanics and in technical mechanics over the past two centuries. The concept of instability has generated a great deal of intensive research, especially during the past few decades.

Like the discrete-time dynamical systems (X,T) that we have been considering up until now in this chapter, a continuous-time dynamical system has a state space X as one of its constituent parts: every element x of the set X is the state of the system. In the case of continuous-time systems, we often refer to the state space as the *phase space* of the system. When we are working with a continuous-time system, we are working with more than just the iterates

$$T^0, T^1, T^2, T^3 \ldots$$

of a single transformation $T : X \to X$ (where T^0 stands for the identity mapping, which we shall denote by id). In a continuous-time dynamical system we have a whole family (T_t) of mappings $T_t : X \to X$ defined for every real number t, and not only for nonnegative integers. Our interpretation of this collection of mappings is that if the system is in state x at any given instant, then after a time t has elapsed, the system will be in the state $T_t(x)$. If we now let a further time interval s elapse, then the system will be in the state $T_s(T_t(x))$. On the other hand, a time interval t followed by a time interval s is precisely a time interval of length $s + t$; and therefore, after these two time intervals have elapsed, the state of the system should be $T_{s+t}(x)$. With this argument in mind, we require a continuous-time dynamical system to satisfy the identity

$$T_{s+t}(x) = T_s(T_t(x))$$

for all real numbers s and t and all elements (states) x in X. This identity can also be written in terms of composition of functions as

(1) $T_{s+t} = T_s \circ T_t.$

The interpretation that we have just given of the symbol $T_t(x)$ also applies to the case $t = 0$, in which the system has undergone no change. Furthermore, if $t > 0$, then we can interpret the state $T_{-t}(x)$ as being the state of the system t units of time ago. In other words,

$$T \circ T_{-t} = T_{t+(-t)} = T_0 = \text{id}.$$

With these comments in mind, we can define a continuous-time dynamical system to be an ordered pair $(X, (T_t))$, where X is a set and (T_t) is a collection of functions from X to X defined for every real number t, and such that the identity (1) is always satisfied.

Given any continuous-time dynamical system, we may, if we like, restrict our attention to $t = 0, 1, \ldots$ and obtain a discrete-time dynamical system (X, T), where $T = T_1$. Note that for every positive integer n,

$$T^n = T_1 \circ \ldots \circ T_1 = T_{1 + \ldots + 1} = T_n.$$

We can think of such a discrete-time system as describing the original continuous-time system in much the same way that a movie camera stores its excerpt of the continuously moving scene that is being photographed. Such a modification of a given continuous-time system is called a *discretization* of that system with respect to time. It is often useful to investigate such discretizations first, and then to go back to the time-continuous original by choosing the time units smaller and smaller.

In a time-continuous dynamical system $(X, (T_t))$, a given point x in X moves through its *orbit* $(T_t(x))$ as t varies continuously through "time." In classical dynamics, the fully fledged system $(X, (T_t))$ in the above sense often appears at a comparatively late stage of a presentation. One usually starts by constructing the orbits of the points x in the space X as the solution curves of what we call *ordinary differential equations*. Only then are the mappings T_t defined in terms of a movement through the orbits in a time interval t. It can be shown that under suitable conditions, these two approaches to continuous-time dynamical systems are equivalent. We mention that an ordinary differential equation can be thought of as providing a direction of travel, an "arrow," at each point. We can think of the act of solving an ordinary differential equation as the act of constructing curves that fit the direction of the arrows at every point. The collection of arrows is known as a *vector field*. The differential equation that determines a dynamical system in the real world is often a statement of some fundamental laws of physics. Thus, for example, the differential equations of planetary motion are statements of Newton's law of gravitation, together with Newton's second law of motion. We have chosen to describe dynamical systems in this section in terms of families (T_t) of mappings in order to avoid the technicalities of calculus.

6.1   Stability

Every dynamical system $(X, (T_t))$ that behaves reasonably from the point of view of physics has the following *continuity property*: Given any positive number t_0, if we change ("perturb") the system from a given state x into a state y that is sufficiently "close" to X, then after any time $t \leq t_0$ has elapsed, the system will be in a state $T_t(y)$ that is close to the state $T_t(x)$. The smaller the perturbation, the closer the state $T_t(y)$ will be to the state $T_t(x)$, and the longer this closeness will last. In other words, if we make y closer to x, then we can make the number t_0 larger. Under special circumstances, it is possible to have this kind of closeness even if we take $t_0 = \infty$. When this happens we can say that if y is close to x, then the state $T_t(y)$ will be close to $T_t(x)$ for every time $t \geq 0$, and we say that the dynamical system is *stable* at the point x. Figure VI.6.1 illustrates these notions of continuity and stability.

Stability problems play a fundamental role in many applications. We shall not go into detail on this subject. Instead, we shall dwell on the following historically famous problem:

Our planetary system is perturbed every now and then by a comet that never returns (a so-called hyperbolic comet). Each time this happens, we are pushed from our previous state x into another state y. Our planetary system is, of course, continuous; therefore, if the comet is small enough, its impact on the system will be small for some limited time (say, a few millennia). However, this knowledge on its own is not enough to make us feel comfortable. It would be nice to know that our system is truly stable in the sense that the fate of our planet does not undergo any radical changes *at any time in the future*. We could, perhaps, interpret this desire for stability as the modern counterpart of the medieval "comet-panic" that was experienced whenever a comet appeared.

There are some extreme possibilities that we need to take into account, in theory at least. On the one hand, the effect of a comet may be so small that our system will return to its original state after a little time. On the other hand, if the comet is large enough,

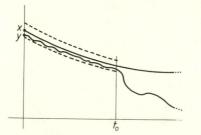
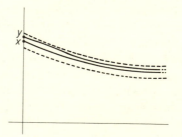

Figure VI.6.1

then it might cause total disintegration of our planetary system. The field known as *perturbation theory*, which describes the effect of such perturbations, celebrated one of its early triumphs in 1846 when the planet Neptune was actually discovered at the point in space at which Urbain Le Verrier (1811–1877) had predicted that a new planet had to exist. Le Verrier based his prediction on the perturbations that had been measured in the orbits of the planets already known at that time.

The problem of the stability of our planetary system and of other similar kinds of systems was one of the most prominent subjects of mathematical research in the middle of the nineteenth century. At that time, such problems were usually treated by the method of expanding solutions into a series. However, this tool was shown to be ineffective when Karl Weierstrass (1815–1897) discovered that the denominators of the fractions that appeared in the terms of those series were liable to be so small that the series could fail to converge. Unless the series converges, it is useless (at least in rigorous mathematics) as a description of anything at all. The objection that Weierstrass raised to the series technique is called the "problem of small denominators" to this day.

This problem motivated Henri Poincaré (1854–1912) to develop some completely new "qualitative" methods for celestial mechanics. These methods led to a breakthrough in the mathematical theory of planetary stability that took place shortly after 1950, and they still pervade mathematical research in this field today. An important part of this theory as it stands today is known as KAM theory, after the initials of the three principal contributors, Andrej Nikolajevic Kolmogorov (1903–1987), Vladimir Igorevic Arnol'd (b. 1937), a student of Kolmogorov's, and Jürgen Moser (b. 1928), a student of Carl Ludwig Siegel's (1896–1981). The main original papers are Kolmogorov [1954, 1957], Arnol'd [1963], and Moser [1966, 1973]. A thorough presentation of this theory at beginning graduate level can be found in Rüssmann [1979]; see also Rüssmann [1983]. The main result of KAM theory can be stated roughly as follows:

> Undisturbed systems of the type of our planetary system evolve like the Kronecker flows that we discussed in Subsection 3.1. Such systems are said to evolve *quasi-periodically*. If these systems undergo sufficiently small perturbations, then they remain quasi-periodic with "high probability," and even close to the quasi-periodicity that prevailed before the perturbation. The smaller the perturbation, the greater the probability of preservation of quasi-periodicity, so that if the perturbation is very small, then the probability that quasi-periodicity will be preserved is very close to the ideal value 1.

The introduction of this probability viewpoint is one of the new features that have been introduced into the field by KAM theory. Using this idea we can roughly say that although chaos is not totally excluded when our planetary system is perturbed, it becomes highly improbable if the perturbation is small enough.

6.2   Instability

If we do not have stability in a given time-continuous dynamical system $(X, (T_t))$ at a given point x, or if we are unable to prove that it exists, then we may simply look into the possibilities that may exist for points that lie near the point x. We may examine these neighboring points by looking into the future as well as looking into the past. The following two extreme patterns of behavior are of particular interest.

(1) We shall say that a point y belongs to the *stable manifold* of a given point x if y becomes a closer and closer neighbor of x in the *future*. More precisely, this condition requires that the distance between $T_t(y)$ and $T_t(x)$ approaches zero as $t \to \infty$.

(2) We shall say that a point y belongs to the *unstable manifold* of x if it belongs to the stable manifold of x in "reversed" time. More precisely, this condition requires that the distance between $T_t(y)$ and $T_t(x)$ approaches zero as $t \to -\infty$.

By enclosing an additional time axis, we can picture these two possibilities in Figure VI.6.2. If we omit the additional time axis from our figure by looking along it into the future, we may visualize a neighborhood of x as in Figure VI.6.3. In this figure we have also shown the behavior of some points that are first attracted by x, and then pushed away again after a while.

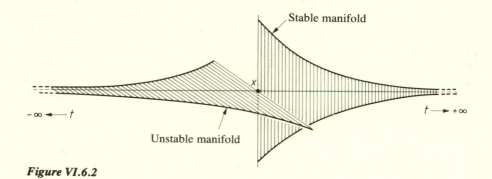

Figure VI.6.2

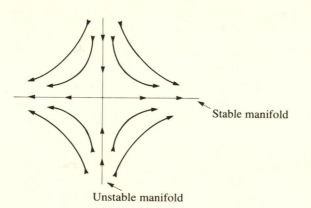

Figure VI.6.3

Not all systems have stable and unstable manifolds, but for those that have them in a reasonable fashion, modern investigations make use of a well-established theory of *unstable* or *chaotic* behavior. Results of this type allow us to understand why systems that are rigorously deterministic still show some *statistical* behavior if we look at them globally. This paradox has haunted statistical mechanics from its beginning around the middle of the nineteenth century.

Investigations of this type are also of particular interest in the theory of *turbulence* (Ruelle-Takens [1971]), and even in *meteorology*, which has recently contributed an especially impressive example of chaotic behavior called the *Lorenz attractor* (see Ruelle [1980]). Researchers are particularly satisfied if they succeed in finding *shifts* as subsystems of classical dynamical systems, since the shift is *the* prototype of random-like behavior. Sometimes this is achieved by proving the existence of horseshoe-like phenomena (Subsection 3.3); see, for example, Moser [1973], and also Jacobs [1978a].

The dynamical system that represents an ideal gas (elastic pellets in a box with reflecting walls) has eluded almost every effort in this direction so far; see Sinai [1963]. A much simplified system in which photons are reflected in two dimensions, the so-called *Sinai Billiard system*, has been successfully investigated. The most important result that has been obtained for this system is its shift representation (Gallavotti-Ornstein [1974]).

Understanding the kind of behavior that leads from one type of evolution in time to a completely different type of evolution is a very general objective of mathematical research. It has led to such results as *catastrophe theory* (Thom [1972], [1984], Zeeman [1977], Arnol'd [1984]) and *bifurcation theory* (see again Arnol'd [1984], and also Iooss-Joseph [1980]). A pioneer in these directions of research was Eberhard Hopf (1902–1983). To nonmathematicians we recommend reading the book by Ekeland [1984].

Literature Cited

[1924] *Alexander, J. W.*, An example of a simply connected surface bounding a region which is not simply connected, Proc. Nat. Acad. Sci. USA *10* (1924), 8–10.

[1927] *Alexander, J. W.*, and *B. G. Briggs*, On types of knotted curves, Ann. of Math. *28* (1927), 562–586.

[1966] *Arbib, M. A.*, Simple self-reproducing automata, Inf. and Control *9* (1966), 177–189.

[1969] *Arbib, M. A.*, Self-reproducing automata—some implications for theoretical biology, in: Waggington, C. H. (ed.): Towards a theoretical biology, vol. 2, Chicago 1969, 204–226.

[1963a] *Arnol'd, V. I.*, Proof of a theorem of A. N. Kolmogorov on the invariance of quasi-periodic motions under small perturbations of the Hamiltonian, Russ. Math. Surveys (Uspekhi) *18* (1963), 9–36.

[1963b] *Arnol'd, V.I.*, Small divisor problems in classical and celestial mechanics, Russian Math. Surveys (Uspekhi) *18* (1963), 85–192.

[1984] *Arnol'd, V. I.*, Catastrophe theory, Berlin–Heidelberg–New York (Springer-Verlag) 1984.

[1951] *Arrow, K.*, Social choice and individual values, New York (Wiley) 1951.

[1979] *Aubin, J. P.*, Mathematical methods of game and economic theory, Amsterdam (North-Holland) 1979.

[1894] *Barlow, W.*, Über die geometrischen Eigenschaften homogener starrer Strukturen und ihre Anwendungen auf Krystalle, Z. Krist. *23* (1984), 1–63.

[1979] *Beck, A.*, Ein Paradoxon: der Hase und die Schildkröte, Sel. Math. V, 1–21, Berlin–Heidelberg–New York (Springer-Verlag) 1979.

[1977] *Beckmann, P.*, A history of π (pi), Boulder/Col. (Golden Press) 1977.

[1987] *Berger, M.* Geometry, 2 vol. Berlin (Springer)1987.

[1982] *Berlekamp, E. R., J. H. Conway* and *R. K. Guy*, Winning Ways, 2 vols., New York (Academic Press) 1982.

[1985] *Beth, Th., D. Jungnickel* and *H. Lenz*, Design Theory, Mannheim–Wien–Zürich (Bibl. Inst.) 1985.

[1952] *Bieberbach, L.*, Theorie der geometrischen Konstruktionen, Basel (Birkhäuser) 1952.

[1931] *Birkhoff, G. D.*, Proof of the ergodic theorem, Proc. Nat. Acad. Sci. USA *17* (1931), 656–660.

[1958] *Black, D.*, The theory of committees and elections, Cambridge (Univ. Press) 1958.

[1916] *Blaschke, W.*, Kreis und Kugel, Leipzig (Veit) 1916, 2. Aufl. Berlin (deGruyter) 1956.

[1974] *Boeckmann, J.*, und *G. Schill*, Knotenstrukturen in der Chemie, Tetrahedron *30* (1974), 1945–1957.

[1896] *Boltzmann, L.*, Entgegnung auf die wärmetheoretische Betrachtung des Herrn Zermelo, Ann. Phys. *57* (1896), 773–784.

[1832] *Bolyai, J.*, Appendix, scientiam spatii absolute veram exhibens, Anhang zu W. Bolyai's Tentamen etc. 1832, Faksimile Budapest (Akad. Kiadó) 1973.

[1983] *Brams, S. I.*, Superior Beings, Heidelberg–New York (Springer-Verlag) 1983.

[1985] *Brams, S. I.*, Superpower Games, New Haven (Yale U.P.) 1985.

[1981] *Brieskorn, E.* and *H. Knörrer*, Ebene algebraische Kurven, Wiesbaden (Vieweg) 1981, eng. transl.: Plane algebraic curves, Basel-Boston (Birkhäuser) 1986.

[1911] *Brouwer, L.E.J.*, Über Abbildungen von Mannigfaltigkeiten, Math. Ann. *71* (1911), 97–115.

[1913] *Brouwer, L.E.J.*, Über den natürlichen Dimensionsbegriff, Crelle's J. *142* (1913), 146–152.

[1978]   *Brown, H., R. Bülow, J. Neubüser, H. Wondratschek* und *H. Zassenhaus*, Crystallographic groups of four-dimensional space, New York (Wiley) 1978.

[1959]   *Burger, E.*, Einführung in die Theorie der Spiele, Berlin (de Gruyter) 1959.

[1878]   *Cantor, G.*, Ein Beitrag zur Mannigfaltigkeitslehre, Crelle's J. *84* (1878), 242–258.

[1883]   *Cantor, G.*, 4 über unendiche lineare Punktmannigfaltigkeiten, Ges. Abh., Berlin (Springerverlag) 1932, 139–246.

[1891]   *Cantor, G.*, Über eine elementare Frage der Mannigfaltigkeitslehre, Jber DMV *1* (1891), 75–78.

[1981]   *Cassels, I.E.W.*, Economics for Mathematicians, Cambridge (Univ. Press) 1981.

[1968]   *Chiu Chang Suan Shu* (Neun Bücher arithemetischer Technik), Ed. K. Vogel, Braunschweig (Vieweg) 1968.

[1966]   *Collatz, L.* and *W. Wetterling*, Optimierungsaufgaben, Berlin–Heidelberg–New York (Springer-Verlag) 1966, engl. trans.: Optimization problems, New York (Springer) 1975.

[1976]   *Conway, J.H.*, On numbers and games (ONAG), New York (Academic Press) 1976.

[1963]   *Coxeter, H.S.M.*, Regular Polytopes, 2nd ed. New York (MacMillan) 1963.

[1969]   *Coxeter, H.S.M.*, Introduction to Geometry, 2nd ed. New York (Wiley) 1969.

[1982]   *Coxeter, H.S.M., P. Du Val, H. T. Flather* and *J. F. Petrie*, The fifty-nine icosahedra, New York–Heidelberg–Berlin (Springer-Verlag) 1982.

1976]   *Critchlow, K.*, Islamic Patterns, London (Thames & Hudson) 1976.

[1951]   *Dantzig, G. B.*, Maximization of a linear function of variables subject to linear inequalities, Cowles Comm. Monogr. *13*, 19–32, New York (Wiley) 1951.

[1986]   *Danzer, L.*, Zur Lösung des Gallai'schen Problems über Kreisscheiben in der euklidischen Ebene, Stud. Sci Math. Hung. (to appear).

[1970]   *Davis, M. D.* Game theory: A non-technical introduction, New York (Basic Books) 1970.

[1976]   *Dawkins, R.*, The selfish gene, Oxford (Univ. Press) 1976.

[1959]   *Debreu, G.*, Theory of value, New Haven (Yale Univ. Press) 1959.

[1888]   *Dedekind, R.*, Was sind und was sollen die Zahlen, Braunschweig (Vieweg) 1888, Neudruck (der 8. Aufl.) 1960, engl. transl.: The nature and meaning of numbers, New York (Dover) 1963.

[1900]   *Dehn, M.*, Über raumgleiche Polyeder, Nachr. Ges. Wiss. Göttingen *1900*, 345–354.

[1902]   *Dehn, M.*, Über den Rauminhalt, Math. Ann. *55* (1902), 465–478.

[1979]   *Dekking, F. M.*, Strongly non-repetitive sequences and progression-free sets, J. Comb.Th. A *27* (1979), 181–185.

[1968]   *Dembowski, P.*, Finite geometries, Berlin–Heidelberg–New York (Springer Verlag) 1968.

[1637]   *Descartes, R.*, La géométrie, 1637.

[1984]   *Driver, R. D.*, Why Math?, Berlin–Heidelberg–New York–Tokyo (Springer Verlag) 1984.

[1876]   *Du Bois–Reymond, P.*, Über die Paradoxien des Infinitärkalküls, Math. Ann. *11* (1876) 149–167.

[1985]   *Dubrovin, B. A., A. T. Fomenko* and *S.P. Novikov*, Modern geometry—methods and applications, 2 vol., Berlin–Heidelberg–New York–Tokyo (Springer) 1984/85.

[1983]   *Ebbinghaus, H.-D., H. Hermes, F. Hirzebruch, M. Koecher, K. Mainzer, A. Prestel, R. Remmert*, Zahlen, Berlin–Heidelberg–New York–Tokyo (Springer-Verlag) 1983.

[1977]   *Edwards, H. M.*, Fermat's last theorem, Berlin–Heidelberg–New York (Springer-Verlag) 1977.

[1907]   *Ehrenfest, P.*, und *T. Ehrenfest*, Über zwei bekannte Einwände gegen das Boltzmannsche H-Theorem, Phys. Z. *8* (1907), 311–314.

[1975]   *Eigen, M.*, und *R. Winkler*, Das Spiel, München (Piper) 1975.

[1915]   *Einstein, A.*, Zur allgemeinen Relativitätstheorie, Sber. Preuß. Akad. Wiss. 1915, 778–786, 799–801, also 831–839, 844–847.

[1916]   *Einstein, A.*, Grundlagen der allgemeinen Relativitätstheorie, Ann. Phys. (4) *48* (1916), 769–822.

[1984]   *Ekeland, I.*, Le calcul, l'imprévu, Paris (Seuil) 1984.

[1976]   *El-Said, I.,* and *A. Parman,* Geometric Concepts in Islamic Art, London (World of Islam Festival Publ. Comp.) 1976.

[1977]   *Engel, A.,* Elementarmathematik vom algorithmischen Standpunkt, Stuttgart (Klett) 1977, engl. transl.: Elementary mathematics from an algorithmic standpoint, Keele, 1984.

[1984]   *Escher, M. E.,* Leben und Werk, Eltville (Rheingauer VG) 1984.

[1891]   *Fedorov, E. S.,* The symmetry of regular systems of figures, Petersburg 1891, reprint, New York (Polycrystal Book Service) 1971.

[1953]   *Fejes Tóth, L.,* Lagerungen in der Ebene, auf der Kugel und im Raum, Berlin–Göttingen–Heidelberg (Springer-Verlag) 1953.

[1964]   *Fejes Tóth, L.,* What the bees know and what they don't know. BAMS *70* (1964), 468–481.

[1965]   *Fejes Tóth, L.,* Reguläre Figuren, Leipzig (Teubner) 1965.

[1978/79]   *Felscher, W.* Naive Mengen und abstrakte Zahlen, 3 vols. Mannheim (Bibl. Inst.) 1978/79.

[1822]   *Feuerbach, Karl,* Eigenschaften einiger merkwürdigen Punkte des geradlinigen Dreiecks, Nürnberg 1822.

[1930]   *Fisher, R. A.,* The genetical theory of natural selection, Oxford (Univ. Press) 1930.

[1956]   *Ford, L. R.,* and *D. R. Fulkerson,* Maximal flow through a network, Can J. Math. *8* (1956), 399–404.

[1962]   *Ford, L. R.,* and *D. R. Fulkerson,* Flows in networks, Princeton (Univ. Press) 1962.

[1980]   *Franklin, J.,* Methods of Mathematical Economics, Berlin–Heidelberg–New York (Springer-Verlag) 1980.

[1903]   *Frege, G.,* Über die Grundlagen der Geometrie, 3 Teile, Jber. DMV *12* (1903), 319–324, 368–375, *15* (1906) 423–430.

[1937]   *Freudenthal, H.,* Zur intuitionistischen Deutung logischer Formeln, Compos. Math. *4* (1937), 112–116.

[1953]   *Freudenthal, H.,* Zur Geschichte der vollständigen Induktion, Arch. Int. Hist. Sci *22* (1953), 17–37.

[1945]   *Fritz, K.v.,* The discovery of incommensurability by Hippasus of Metapontum, Ann. of Math. *46* (1945), 242–264.

[1974]   *Gallavotti, G.,* and *D. S. Ornstein,* Billiards and Bernoulli schemes, Comm. Math. Phys. *38* (1974), 83–101.

[1796]   *Gauss, C. F.* (publ. through E.A.W. Zimmermann), Zur Kreistheilung, Intelligenzblatt d. allg. Literaturzeitung 1.6.1796, S. 554, Werke X, 3; cf. S. 120–126.

[1936]   *Gentzen, G.,* Die Widerspruchsfreiheit der reinen Zahlentheorie, Math. Ann *112* (1936), 493–565.

[1825]   *Gergonne, J. D.,* Ann. de Math. *16* (1825), 209–231.

[1973]   *Gibbard, A.,* Manipulation of voting schemes: a general result, Econometrica *41* (1973), 587–601.

[1931]   *Gödel, K.,* Über formal unentscheidbare Sätze der Principia Mathematica und verwandter Systeme I., Monatsh. Math. Phys. *38* (1931), 173–198.

[1981]   *Goldstein, S.,* Entropy increase in dynamical systems, Isr. J. Math. *38* (1981), 241–256.

[1964]   *Gottschalk, W. H.,* A characterization of the Morse minimal set, Proc. AMS *15* (1964), 70–74.

[1980]   *Graham, R., B. Rothschild* and *J. Spencer,* Ramsey Theory, New York (Wiley) 1980.

[1967]   *Greub, W. H.,* Linear algebra, Berlin–New York (Springer) 1967.

[1983]   *Grünbaum, B.,* and *G. L. Shephard,* Tilings, patterns, fabrics and related topics in discrete geometry, Jber. DMV *85* (1983), 1–32.

[1955]   *Guggenbuhl, L.,* Karl Wilhelm Feuerbach, mathematician, The Sci. Monthly *81* (1955).

[1957]   *Hadwiger, H.,* Vorlesungen über Inhalt, Oberfläche und Isoperimetrie, Berlin–Göttingen–Heidelberg (Springer-Verlag) 1957.

[1950]   *Halmos, P. R.,* Measure theory, New York (Nostrand) 1950.

[1935]   *Hall, P.*, On representations of subsets, J. London Math. Soc. *10* (1935), 26–30.

[1950]   *Halmos, P. R.*, and *H. Vaughan*, The marriage problem, Amer. J. Math. *72* (1950), 214–215.

[1954]   *Hardy, G. H.*, and *E. M. Wright*, An introduction to the theory of numbers, 4th ed. Oxford (Univ. Press) 1960.

[1944]   *Hedlund, G.*, and *M. Morse*, Unending chess, Symbolic dynamics and a problem in semigroups. Duke J. Math. *11* (1944), 1–7.

[1968]   *Heinisch, K. I.*, Kaiser Friedrich II. in Briefen und Berichten seiner Zeit, Darmstadt (Wiss. Buchges.) 1968.

[1830]   *Hessel, J.F.C.*, Krystallometrie oder Krystallonomie und Krystallographie, Gehler's physikalisches Wörterbuch Bd. 5 (1830).

[1831]   *Hessel, J.F.C.*, Kristallometrie, Leipzig 1831.

[1891]   *Hilbert, D.*, Über die stetige Abbildung einer Linie auf ein Flächenstück, Math. Ann. *38* (1891), 459–460.

[1899]   *Hilbert, D.*, Grundlagen der Geometrie, Leipzig (Teubner) 1899.

[1900]   *Hilbert D.*, Mathematische Probleme, Göttinger Nachr. *1900*, 253–297, Ges. Abh. Bd. III, 290–329.

[1985]   *Hildebrandt, S.*, and *A. Tromba*, Mathematics and optimal form, New York (Freeman) 1985.

[1975]   *Hildenbrand, W.*, und *K. Hildenbrand*, Lineare ökonomische Modelle, Berlin–Heidelberg–New York (Springer-Verlag) 1975.

[1979]   *Hlawka, E.*, Theorie der Gleichverteilung, Mannheim (BI) 1979.

[1988]   *Hofbauer, J.*, and *K. Sigmund*, The theory of evolution and dynamical systems, transl. from the German 1984 ed., Cambridge (Univ. Press) 1988.

[1980]   *Iooss, G.*, and *D. D. Joseph*, Elementary stability and bifurcation theory, Berlin–Heidelberg–Tokyo (Springer-Verlag) 1980.

[1969a]   *Jacobs, K.*, Maschinenerzeugte 0-1-Folgen, Selecta Math. I, Berlin–Heidelberg–New York (Springer-Verlag) 1969, 1–27.

[1969b]   *Jacobs, K.*, Der Heiratssatz, Sel. Math. I, Berlin–Heidelberg–New York (Springer-Verlag) 1969, 103–141.

[1972]   *Jacobs, K.*, Einige Grundbegriffe der topologischen Dynamik, Sel. Math. IV (1972), 1–30.

[1978a]   *Jacobs, K.*, Stochastics and mechanics, Bull. Inst. Math. Acad. Sinica *6* (1978), 429–456.

[1983a]   *Jacobs, K.*, Einführung in die Kombinatorik, Berlin (de Gruyter) 1983.

[1983b]   *Jacobs, K.*, Arithmetische Progressionen, Jber. DMV *85* (1983), 55–65.

[1969]   *Jacobs, K.*, and *M. Keane*, 0-1-Sequences of Toeplitz type, ZfW *13* (1969), 123–131.

[1984]   *Jacobs, K.*, and *H. Utz*, Erlangen programs, Math. Intell. *6* (1984), 79.

[1968]   *Jessen, B.*, The algebra of polyhedra and the Dehn-Sydler theorem, Math. Scand. *23* (1968), 241–256.

[1868]   *Jones, O.*, The grammar of ornament, London (Quaritch) 1868.

[1941]   *Kakutani, S.*, A generalization of Brouwer's fixed point theorem, Duke J. Math. *8* (1941), 457–459.

[1968]   *Keane, M.*, Generalized Morse sequences, Z. f. Warsch. *10* (1968), 335–353.

[1955]   *Kelley, J. L.*, General topology, New York (van Nostrand) 1955.

[1619]   *Kepler, J.*, Harmonia Mundi, 1619.

[1977]   *Kleber, W.*, Einführung in die Kristallographie, Berlin (Verlag Technik) 1977.

[1872]   *Klein, F.*, Vergleichende Betrachtungen über neuere geometrische Forschungen, 48 S., Erlangen (Deichert) 1872, augmented Text in Math. Ann.*43* (1893) and Ges. Abh. Bd. I.

[1882]   *Klein, F.*, Über Riemanns Theorie der algebraischen Funktionen und ihrer Integrale, Leipzig (Teubner) 1882, Ges. Abh. Bd. *3*, 499–586.

[1884]   *Klein, F.*, Vorlesungen über das Ikosaeder, Leipzig 1884.

[1979]   *Kline, M.*, (ed.), Mathematics, an introduction to its spirit and use. San Francisco (Freeman) 1979.

[1973] *Klingenberg, W.*, Eine Vorlesung über Differentialgeometrie, Berlin–Heidelberg–New York (Springer-Verlag) 1973.

[1968] *Knuth, D.*, The art of Computer Programming, 3 vols. Reading/Mass. (Addison-Wesley) 1968ff.

[1983] *Koecher, M.*, Lineare Algebra und analytische Geometrie, Berlin–Heidelberg–New York–Tokyo (Springer-Verlag) 1983.

[1916] *König, D.*, Über Graphen und ihre Anwendungen, Math. Ann. *77* (1916), 453–465.

[1985] *König, H.*, und *M. Neumann*, Mathematische Wirtschaftstheorie, Königstein (Hain/Athenäum) 1985.

[1982] *Kötter, R.*, General equilibrium theory—an empirical theory ?, Stud. Contemp. Ec. *2* (1982), 103–117.

[1954] *Kolmogorov, A. N.*, On conservation of conditionally periodic motions for a small change in Hamilton's function. Doklady Akad. Nauk SSSR (N.S.) *98*, 527–530 (1954) (Russian).

[1957] *Kolmogorov, A. N.*, Théorie générale des systèmes dynamiques et mécanique classique. Proceedings of the International Congress of Mathematicians, Amsterdam, 1954, vol. 1, pp. 315–333. Erven P. Noordhoff N. V., Groningen: North Holland Publishing Co., Amsterdam.

[1971] *Kornai, J.*, Anti-Equilibrium, Amsterdam (North Holland) 1971.

[1970] *Kowalsky, H.*, Lineare Algebra, 5th ed., Berlin (de Gruyter) 1970.

[1884] *Kronecker, L.*, Die Periodensysteme von Funktionen reeller Variablen. Näherungsweise ganzzahlige Auflösung linearer Gleichungen, Kgl. Preuss. Akad. Wiss. *1884*, 1071–1080, 1179–1193, 1271–1299; Werke Bd. III, 31–46, 47–109,

[1950] *Kuhn, W.*, Extensive games, Proc. Nat. Acad. Sci. USA *36* (1950), 570–576.

[1980] *Lam Lay-Yong*, The connection between the Pascal Triangle and the solution of numerical equations of any degree, Hist. Math. *7* (1980), 407–424.

[1930] *Landau, E.*, Grundlagen der Analysis, Leipzig (Akad. Verlagsgesell.) 1930, engl. transl.: Foundations of Analysis, New York (Chelsea) 1951.

[1953] *Lietzmann*, Der pythagoreische Lehrsatz, Stuttgart (Teubner) 1953.

[1882] *Lindemann, F.*, Über die Zahl π, Math. Ann. *20* (1882), 213–225.

[1829] *Lobacevski, N. I.*, Über die Anfangsgründe der Geometrie (russ.) Kaz. Vestnik (Kasaner Bote) 1829.

[1940] *Loomis, E. S.*, The Pythagorean Proposition, 2nd ed., Washington D.C. (Natl. Counc. of Teachers of Math.) 1972.

[1935] *Maak, W.*, Eine neue Definition der fastperiodischen Funktionen, Abh. Math. Sem. Hamburg *11* (1935), 240–244.

[1970] *MacLane, S.*, Categories for the working mathematician, New York (Springer), 1970.

[1930] *Maennchen, Ph.*, Gauss als Zahlenrechner, Gauss Werke X, 2, Nr. 6 Berlin (Springer-Verlag) 1930.

[1937] *Mahler, K.*, Arithmetische Eigenschaften einer Klasse von Dezimalbrüchen, Proc. Acad. Wet. Amsterdam *40* (1937), 421–428.

[1977] *Mandelbrot, B. M.*, Fractals: Form, chance and dimension, San Francisco (Freeman) 1977.

[1967] *Massey, W. S.*, Algebraic topology: An introduction, Berlin–Heidelberg–New York (Springer-Verlag) 1967.

[1909] *Minkowski, H.*, Theorie der konvexen Körper, insbesonders Begründung ihres Oberflächenbegriffs; aus dem Nachlaß veröff. in Ges. Abhandlungen Bd. II, 131–229, Leipzig (Teubner) 1911.

[1977] *Moise, E. E.*, Geometric topology in dimensions 2 and 3, Berlin–Heidelberg–New York (Springer-Verlag) 1977.

[1962] *Moore, E. F.*, Machine models of self-reproduction, Proc. Symp. Appl. Math. *14* (1962), 17–33.

[1921] *Morse, M.*, Recurrent geodesics on a surface of negative curvature, TAMS *22* (1921), 84–100.

[1966]   *Moser, J.*, On the theory of quasi-periodic motions, Lectures held at Stanford Univ. 1965, SIAM Review *8* (1966), 145–172.

[1973]   *Moser, J.*, Stable and random motions in dynamical systems with special emphasis on celestial mechanics, Princeton (Univ. Press) 1973.

[1950]   *Nash, J.*, Equilibrium points in n-person games, Proc. Nat. Acad. Sci. USA *36* (1950), 48–49.

[1951]   *Nash, J.*, Non-cooperative games, Ann. of Math (2) *54* (1951), 286–295.

[1928]   *Nelson, L.*, Kritische Philosophie und mathematische Axiomatik, Unterrichtsbl. f. Math. u. Naturw. *34* (1928), 108–115, 136–142, abgedruckt in: Thiel, C., Erkenntnistheoretische Grundlagen der Mathematik, Hildesheim (Gerstenberg) 1982.

[1961]   *Nemeth, L.*, A két Bolyai (Die beiden Bolyai), Budapest (Szépirodalmi Könyvkiadó) 1961, Uraufführung 20.4.61 im Katona-Jószef-Theater Budapest.

[1928]   *Neumann, J.v.*, Zur Theorie der Gesellschaftsspiele, Math. Ann. *100* (1928), 295–320.

[1975]   *Nöbeling, G.*, Einführung in die nichteuklidischen Geometrien der Ebene, Berlin (de Gruyter) 1975.

[1968]   *Owen, G.*, Game theory, Philadelphia (Saunders) 1968.

[1977]   *Paris, J.*, and *L. Harrington*, A mathematical incompleteness in Peano arithmetic, in: Handbook of Math. Logik (ed. Barwise), 1133–1142, Groningen (North-Holland) 1977.

[1665]   *Pascal, B.*, Traite du triangle arithmétique etc. Paris (Desprez) 1665.

[1882]   *Pasch, M.*, Vorlesungen über neuere Geometrie, Leipzig (Teubner) 1882, 2nd ed., with an appendix by Max Dehn, Berlin (Springer) 1926.

[1889]   *Peano, G.*, Arithmetices principia novo modo exposita, Torino (Bocca) 1889, in Op. Sc. II (1958).

[1890]   *Peano, G.*, Sur une sourbe qui remplit toute une aire plane, Math. Ann. *36* (1890), 157–160.

[1978]   *Peleg, B.*, Consistent voting schemes, Econometrica *46* (1978), 153–161.

[1984]   *Peleg, B.*, Game theoretic analysis of voting in committees, Cambridge (Univ. Press) 1984.

[1933]   *Perron, O.*, Eine neue Winkeldreiteilung des Schneidermeisters Kopf, Sber. Bayr. Akad. Wiss. *1933*, 439–445.

[1962]   *Perron, O.*, Nichteuklidische Elementargeometrie der Ebene, Stuttgart (Teubner) 1962.

[1983]   *Petersen, K.*, Ergodic Theory, Cambridge (Univ. Press) 1983.

[1955]   *Pickert, G.*, Projektive Ebenen, Berlin–Göttingen–Heidelberg (Springer-Verlag) 1955.

[1983]   *Pötters, W.*, La natura e L'origine del sonetto. Una nuova teoria, Firenze, Olschki 1983, I, 71–78.

[1890]   *Poincaré, H.*, Sur le problème des trois corps et les équations de la dynamique, Acta Math. *13* (1890), 1–271.

[1895]   *Poincaré, H.*, Analysis situs, J. Ec. Polyt. *1* (1895), 1–121, Oeuvres vol. *6*, 193–288.

[1924]   *Pólya, G.*, Über die Analogie der Kristallsymmetrie in der Ebene, Z. Krist. Min. *60* (1924), 278–282.

[1962]   *Pólya, G.*, Mathematik und plausibles Schließen, 2 Bde. Basel–Stuttgart (Birkhäuser) 1962.

[1970]   *Rabinovitch, N. L.*, Rabbi Levi Ben Gershon and the origins of mathematical induction, Arch. Hist. Ex. Sci. *6* (1970), 237–248.

[1930]   *Ramsey, F. P.*, On a problem of formal logic, Proc. London Math. Soc. *30* (1930), 264–286.

[1932]   *Reidemeister, K.*, Knotentheorie, Berlin (Springer-Verlag) 1932.

[1961]   *Rempel, L. I.*, Architekturnij Ornament Uzbekistana, Taschkent (Gos. Izd.) 1961.

[1979]   *Ribenboim, P.*, 13 Lectures on Fermat's last theorem, Berlin–Heidelberg–New York (Springer-Verlag) 1979.

[1854]   *Riemann, B.*, Über die Hypothesen, welche der Geometrie zugrundeliegen, Werke (Leipzig/Teubner) 1876, 254–269.

[1525]   *Riese, A.*, Rechnung auf der Linien etc., Frankfurt (Egenolph) 1525, Faksimile Hannover (Vincentz) 1978.

[1979] *Roberts, F. S.*, Measurement theory, Reading/Mass. (Addison-Wesley) 1979.

[1972] *Rosenmüller, J.*, Konjunkturschwankungen, Sel. Math. IV, 143–173, Berlin–Heidel-berg–New York (Springer-Verlag) 1972.

[1985] *Rowe, D. E.*, Klein's "Erlanger Antrittsrede," Hist. Math. *12* (1985), 123–141.

[1971] *Ruelle, D.*, and *F. Takens*, On the nature of turbulence, Comm. Math. Phys. *26* (1971), 167–192.

[1980] *Ruelle, D.*, Strange attractors, Math. Intell. *2* (1980), 126–137.

[1979] *Rüssmann, H.*, Konvergente Reihenentwicklungen in der Störungstheorie der Himmels-mechanik, Sel. Math. V, 93–260, Berlin–Heidelberg–New York (Springer-Verlag) 1979.

[1983] *Rüssmann, H.*, Das Werk C. L. Siegels in der Himmelsmechanik, Jber DMV *85* (1983), 174–200.

[1963] *Ryser, H. I.*, Combinatorial mathematics, New York (Wiley) 1963.

[1977] *Sachs, R. K.*, and *H. Wu*, General relativity for mathematicians, Berlin–Heidelberg–New York (Springer-Verlag) 1977.

[1975] *Satterthwaite, M.A.*, Strategy proofness and Arrow's conditions: existence and corre-spondence theorems for voting procedures and social welfare functions, J. Ec. Th. *10* (1975), 187–217.

[1973] *Scarf, H.*, The computation of economic equilibria, New Haven (Yale Univ. Press) 1973.

[1923] *Schmidt, E.*, Über den Jordan'schen Kurvensatz, S. Ber. preuss. Akad. Wiss. *28* (1923), 318–329.

[1891] *Schoenflies, A.*, Krystallsysteme und Krystallstruktur, Leipzig (Teubner) 1891, Nach-druck Berlin–Heidelberg–New York (Springer-Verlag) 1983.

[1962] *Seidenberg, A.*, The ritual origin of geometry, Arch. Hist. Ex. Sci. *1* (1962), 488–527.

[1978] *Seidenberg, A.*, The origin of mathematics, Arch. Hist. Ex. Sci. *18* (1978), 301–342.

[1934] *Seifert, H.*, und *W. Threlfall*, Lehrbuch der Topologie, Leipzig (Teubner) 1934.

[1980] *Selten, R.*, A note on evolutionary stable strategies in asymmetric animal conflicts, J. Theor. Biol. *84* (1980), 93–101.

[1982] *Selten, R.*, Einführung in die Theorie der Spiele mit unvollständiger Information, Schr. Ver. Soc. Pol. *126* (1982), 81–147.

[1974] *Shafarevic, I. R.*, Basic algebraic geometry, Berlin (Springer) 1974.

[1974] *Shubnikov, A. V.*, and *V. A. Koptsik*, Symmetry in science and art, New York (Plenum) 1974.

[1963] *Sinai, Ya. G.*, On the foundations of the ergodic hypothesis for a dynamical system of statistical mechanics, Dokl. Akad. Nauk SSSR *153* (1963), 1261–1264; Engl.: in Soviet Math. Dokl. *4* (1963), 1818–1822.

[1970] *Sinai, Ya. G.*, Dynamical systems with elastic reflections, Uspekhi Mat. Nauk *25* (1970), 141–192; Engl.: in Russian Math. Surveys *25* (1970), 137–189.

[1965] *Smale, S.*, Diffeomorphisms with many periodic points, Diff. and Comb. Top., Princeton (Univ. Press) 1965, 63–80.

[1982] *Smith, J. M.*, Evolution and the theory of games, Cambridge (Univ. Press) 1982.

[1928] *Sperner, E.*, Neuer Beweis für die Invarianz der Dimensionszahl und des Gebietes, Abh. math. Sem. Hamburg *6* (1928), 265–272.

[1952] *Spoerri, Th.*, Genie und Krankheit; eine psychopathologische Untersuchung der Familie Feuerbach, Basel (Karger) 1952.

[1970] *Steen, L. A.*, and *I. A. Seebach*, Counterexamples in topology, New York (HRW) 1970.

[1980] *Straffin, P. D.*, Topics in the theory of voting, Basel–Stuttgart–Boston (Birkhäuser) 1980.

[1969] *Strassen, U.*, Gaussian elimination is not optimal, Numer. Math. *13* (1969), 354–356.

[1986] *Strubecker, K.*, Wilhelm Blaschkes mathematisches Werk, JBer. DMV *88* (1986), 146–157.

[1952/65] *Sydler, I. P.*, Sur les conditions nécessaires pour l'équivalence des polyèdres eucli-déennes, Elem. d. Math. *7* (1952), 49–53; Sur L'équivalence des plyèdres à dièdres ra-tionnels, Elem. d. Math. *8* (1953), 75–79; Conditions nécessaires et suffisantes pour l'é-quivalence des polyèdres de l'espace euclidien à trois dimensions, Comm. Math. Helv. *40* (1965), 43–80.

[1975]   *Szemerédi, E.*, On sets of integers containing no k in arithmetic progression, Acta Arith. 27 (1975), 199–245.

[1972]   *Thom, R.*, Stabilité structurelle et morphogenèse, Reading/Mass. (Benjamin) 1972.

[1984]   *Thom, R.*, Paraboles et Catastrophes, Paris (Flammarion) 1984.

[1904]   *Thue, A.*, Über unendliche Zeichenreihen, Christiania Vid. Selsk. Skr. *1906* T. 7, 22 S. Lex. 8⁰, Werke 139–158.

[1928]   *Toeplitz, O.*, Beispiele zur Theorie der fastperiodischen Funktionen, Math. Ann. *98* (1928), 281–295.

[1953]   *Trost, E.*, Primzahlen, Basel–Stuttgart (Birkhäuser) 1953.

[1936]   *Voderberg, H.*, Zur Zerlegung eines ebenen Bereiches in kongruente, Jber. DMV *46* (1936), 229–231.

[1937]   *Voderberg, H.*, Zur Zerlegung der Ebene in kongruente Bereich in Form einer Spirale, Jber. DMV *47* (1937), 159–160.

[1927]   *Waerden, B. L. van der*, Beweis einer Baudet'schen Vermutung, Nieuw Ark. Wisk. *15* (1927), 212–216.

[1970]   *Waerden, B. L. van der*, Algebra, vol. 1, Engl.: New York (Ungar) 1970.

[1983]   *Waerden, B. L. van der*, Geometry and algebra in ancient civilizations, Berlin–Heidelberg–New York–Tokyo (Springer-Verlag) 1983.

[1985]   *Waerden, B. L. van der*, A history of algebra from al-Khwarizmi to Emmy Noether, Berlin–Heidelberg–New York–Tokyo (Springer-Verlag) 1985.

[1985]   *Wagon, S.*, Is π normal? Math. Intell. *7* (1985), 65–67.

[1986]   *Wagon, S.*, Fermat's last theorem, Math. Intell. *8* (1986), 59–61.

[1856]   *Waltershausen, Sartorius von*, Gauss zum Gedächtnis, Leipzig 1856, Reprint Wiesbaden 1965.

[1837]   *Wantzel, P. L.*, Recherches sur les moyens de reconnaître si un Problème de Géométrie peut se resoudre avec la règle et le compas, J. de math. pures et appl. *2* (1837), 366–372.

[1916]   *Weyl, H.*, Über die Gleichverteilung von Zahlen mod Eins, Math. Ann. *77* (1916), 313–352.

[1949]   *Weyl, H.*, Almost periodic vector sets in a metric vector space, Amer. J. Math. *71* (1949), 178–205.

[1952]   *Weyl, H.*, Symmetry, Princeton (Univ. Press) 1952.

[1977]   *Wickler, W.*, und *U. Seibt*, Das Prinzip Eigennutz, Hamburg (Hoffmann & Campe) 1977.

[1921]   *Wittgenstein, L.*, Tractatus logico-philosophicus, Ann. der Naturphil. 14, 1921, Engl. version: London 1922.

[1948]   *Zassenhaus, H.*, Über einen Algorithmus zur Bestimmung der Raumgruppen, Comm. Math. Helv. *21* (1948), 117–141.

[1977]   *Zeeman, E. C.*, Catastrophe Theory, selected papers, Reading/Mass. (Addison-Wesley) 1977.

[1984]   *Zeh, H. D.*, Die Physik der Zeitrichtung, Lect. Notes Phys. *200* (1984).

[1896]   *Zermelo, E.*, Über einen Satz der Dynamik und der mechanischen Wärmetheorie, Ann. Phys. *57* (1896), 485–494, Über die mechanische Erklärung irreversibler Vorgänge, Ann. Phys. *59* (1896), 793–801.

Index

The selection of entries in this index follows the principle "better too much than too little" in order to facilitate recovery by random association of words, and in order, too, to give a glimpse into the daily verbiage of mathematicians.